U0299568

[英国] 菲利普·鲍尔 著　刘熙 译

分子

牛津通识读本·

Molecules

A Very Short Introduction

译林出版社

图书在版编目（CIP）数据

分子/（英）菲利普·鲍尔（Philip Ball）著，刘熙译．—南京：译林出版社，2017.2（2023.5重印）
（牛津通识读本）
书名原文：Molecules: A Very Short Introduction
ISBN 978-7-5447-6547-3

Ⅰ.①分… Ⅱ.①菲… ②刘… Ⅲ.①分子生物学 - 研究 Ⅳ.①Q7

中国版本图书馆 CIP 数据核字（2016）第 181035 号

著作权合同登记号 图字：10-2014-197 号

分子 ［英国］菲利普·鲍尔 / 著 刘 熙 / 译

责任编辑 许 丹 何本国
责任印制 董 虎

原文出版 Oxford University Press, 2001
出版发行 译林出版社
地 址 南京市湖南路 1 号 A 楼
邮 箱 yilin@yilin.com
网 址 www.yilin.com
市场热线 025-86633278
排 版 南京展望文化发展有限公司
印 刷 江苏凤凰通达印刷有限公司
开 本 890 毫米 × 1260 毫米 1/32
印 张 10.75
插 页 4
版 次 2017 年 2 月第 1 版
印 次 2023 年 5 月第 7 次印刷
书 号 ISBN 978-7-5447-6547-3
定 价 39.00 元

序言

万立骏

写好一部科普科幻作品绝非易事，写好一本科学通识读本也见功力！因而，我们有对《十万个为什么》的怀念和感恩，有对“雨果奖”和“安万特奖”的崇拜，呼唤和期待更多的好作品问世！

今天呈现在诸君面前的《分子》由英国著名科学作家菲利普·鲍尔原著，为“牛津通识读本”系列之一，也是科普作品中的上品。作者菲利普·鲍尔是科普作品大家，多年担任著名学术杂志*Nature*（《自然》）编辑，兼任*Chemistry World*（《化学世界》）、*Nature Materials*（《自然材料》）、*BBC Future*（英国广播公司《未来》节目）等专栏作家。2005年他所著*Critical Mass: How One Thing Leads to Another*（中译本为《预知社会：群体行为的内在法则》）一书获“安万特奖”。

众所周知，分子是保持物质化学性质的最小微粒，由原子、分子构成的物质世界多姿多彩，活力无限！“分子”不仅是“化学”的代名词，也是材料、生命等诸多研究领域的研究对象和诸多功能的执行单元。不同学科领域的科学家对分子的理解、认识和研究角度或许不同，正是这种多层次和多角度的理解使得人们对自然界和对生命体的认识不断加深、不断发现、不断创造。本书的特点，也是精妙之处，是从不同视角介绍了化学、生命、材料等领域的分子研究特点、研究热点和研究成果，又将分子反应、分子材料和分子体系的功能与生命现象、分子机器、分子内/间的信息传递和存储、分子计算技术融会贯通，呈现给读者一个魅力无穷、前景无限的分子世界。作者以通俗的语言、形象的比喻、生动的实例告诉读者什么是分子，分子有什么用途、能做什么，分子科学向什么方向发展等等，写出了一本难得的科学通识作品。

当今时代，新的分析技术不断涌现，“看到”分子已不再是难事；各路科学家攻坚克难，分子层次的科学研究成果精彩纷呈。本书图文并茂、内容丰富、语言优美易懂，英文版自出版以来，深受广大读者的欢迎和好评。此书现在中国出版恰逢其时，一定会吸引更多的青少年朋友对分子科学产生兴趣，激发众多青年才俊的科学创新创造能力，推动分子科学研究的更大发展！我愿意将此书推荐给各位：开卷必有益，或许会获益终身！

目 录

前 言

亚历山大·芬德莱在1916年写下《化学为人服务》这部作品，那时的化学界正迫切需要向全世界宣传化学带来的恩惠。90多年后的今天，我们再写化学，似乎不该再负有那样的重担了。但事实不然。虽然化学的技艺为社会做出了革命性的贡献——单是医学中的化学疗法延长了人类寿命这一点就是明证，可芬德莱那时说的话至今言犹在耳：

> 有的人仅凭一己之力扩大了产业范围，提高了大众的劳动效率，而对科学蒙昧无知的社会大众却疑虑重重，横眉冷对。

这样严峻的声音我们今天依然能够听到，就在化工界及支持者们在面对公众的非难和指责而辩解时。其中一个问题是，化学之善，一旦进入市场就被人们看作天经地义，而化学之恶，却会长久地印在人们的脑海中。同时不可否认的是，化工企业及政府面对沙利度胺[①]和博帕尔[②]等悲剧事件时，以及面对灾难性的臭氧空洞问题时推脱责任，极大地损害了他们为自己辩护的可信度。

于是在进入21世纪时，社会上流行着这样一种想法："化学的"或者"合成的"是糟的，而"天然的"则是好的。

对于这种想法，传统的纠正方式是罗列化学给我们带来的种种好东西。这样罗列出的清单的确很长，即便那些将化工妖魔化的人想必也在享用不少化工产品。可是我相信，我们所需要的已经不再是"化学为人服务"了。一方面来讲，这会更让公众感觉铁板一块的科学和技术共同体抱成一团促进自己的事业。而在外人看来，任何文化圈子都像是铁板一块，因而也都有潜在的威胁。事实上，化学家之间也会就该不该禁止或限制某种产品而爆发激烈的争吵，在有的化学家为军事机构工作时别的化学家则在门外抗议他们——倘若有一天，公众能够更多地了解到这些，那该多好。或许那时我们才能平和地将科学看作正常的人类活动。

然而，从另一方面来讲，其实化学也并不单纯是被人类驯服、召唤来服务的奴仆。化学**造就**我们人类，化学创造了自然界的一切。"化学"和"合成"给人们的负面印象一时还很难消除，但"分子"暂时还没有沾染到这样的色彩。通过理解我们自身在分子层面的本质，也许我们才能开始欣赏化学所提供的一切，并且去领悟为何有些物质（既包括天然的也包括人工的）毒害

① 1950年代问世的药物，又名反应停，可用于治疗妊娠反应。但数年后人们才发现，药效成分的对映异构体具有致畸作用，造成了上万名胎儿的先天畸形。——译注

② 1984年12月3日凌晨，印度博帕尔市，美国联合碳化公司的农药厂发生异氰酸酯泄漏，一个月内逾两万人中毒死亡，几十万人受伤、致残或受到后续影响。——译注

我们而有些物质治愈我们。

正因为如此，我要冒着被一些化学家反对之虞撰写这样一本书，名义上是关于分子的简介，却很大程度上只着重于生命的分子，即生物化学。我努力想说明，那些控制着我们人体的分子层面的过程，其实与化学家——我更愿意称为“分子科学家”——想要创造的区别不大。实际上两者的界限正变得越来越模糊：我们已经在技术领域使用天然的分子，同时又在用合成的分子来保存那些我们看作“天然”的东西。

在讲述这些分子故事的时候，很多专家的建议让我受益良多，这其中包括克格格·比森、保罗·卡尔弗特、乔·霍华德、埃里克·库尔、汤姆·摩尔和乔纳森·斯科里。我在此向他们表示诚挚的感谢。 ii

本书自始至终都是牛津大学出版社通识读本系列的一员。我非常感谢谢利·考克斯对它充满信心，给它一个机会开启自己的生命之旅。

菲利普·鲍尔

2001年1月于伦敦 iii

致 谢

作者和出版社向各方授权使用下列版权保护材料表示由衷的感谢：

弗兰·奥布莱恩《达尔基档案》节选（Copyright © The Estate of the Late Brian O'Nolan），由A.M.希斯有限公司代理授权。

普里莫·莱维《元素周期表》节选，雷蒙德·罗森塔尔译成英文（Copyright © Schocken Books 1984），由兰登书屋旗下肖肯出版社授权使用。

托马斯·品钦《万有引力之虹》节选（Copyright © Thomas Pynchon 1973），由梅兰妮·杰克逊有限责任公司授权翻印。

第一章

无形世界的工程师：制造分子

警官叫来服务员，给自己点了大麦啤酒，又给他的朋友点了一小瓶“那玩意”。然后诡秘地把身子靠了过来。

“你可曾发现过，或者听说过分子吗？”他问道。

“当然了。”

“那你要是知道分子理论正在达尔基区横行，会不会感到很惊讶？”

“嗯……大概吧。”

“它正在肆虐，”他接着说道，“有一半的人都受到了侵袭，简直比天花还糟。”

“难道不能让医生或者教师来解决吗？还是你觉得这应该是各家当家人的责任？”

“这从头到尾，完完全全，都是郡议会的事！”他言语颇为激烈。

“看来真是个麻烦。”

关于分子短而又短的简介其实早就有人写过了，而且写得比我远为巧妙。弗兰·奥布赖恩[1]总喜欢伴着一杯浓烈的健力士啤酒[2]来给我们端出他的博学，仿佛是在谈论土豆田或是都柏林市郊糟糕的路况。在都柏林主干道上的大都会酒店，福特雷警官正在和米克分享他的智慧，而我们也能从中收获些东西：

“你年轻的时候研究过分子理论吗？”他问道。米克说没有，没怎么研究过。

“这玩意可真是肆无忌惮、变本加厉，”他严肃地讲，“不过还是让我来告诉你它有多严重。一切事物都是由事物自身的那种小小的分子所组成的，这些分子飞来飞去，有转圆圈的，有绕弧线的，有直着飞的，轨迹各种各样，无穷无尽。它们从来都不会停着不动，而总是不断地转来转去，到处乱闯，一刻都没停过。你明白了吗？这样的分子？”

“我觉得算明白了。”

“这就像是20只活蹦乱跳的小妖精在平整的墓石上上蹿下跳。用羊来打个比方。究竟羊是什么东西呢？其实，所谓羊，只不过是数以百万计的微小的羊分子在身体里转来转去、上下翻飞而已。”

究竟羊是什么东西呢？这个貌似简单（在一层层的伪装之

---

① 弗兰·奥布赖恩（1911—1966），爱尔兰作家。——译注

② 健力士（Guinness），味道较浓烈的黑啤酒，又译作吉尼斯，即吉尼斯世界纪录的创办者。——译注

下）的问题却足以让科学家困惑数百年之久，而且还会在未来的很长时间里继续让他们困惑下去。分子科学给出了一种分层次的回答：这种科学所考虑的就是那“组成羊的数以百万计的某种微小的羊质”，亦即分子。一只羊是很多种分子组成的混合物，实际上包含着数万种不同的分子。其中的很多种分子不仅包含在羊里面，在人身体里、在青草中，甚至在天空和大海中也都能找到。

但科学不能仅停留于此，而要寻求更深层的理解。羊的分子不又是由原子组成的吗？原子不又是由电子、质子等亚原子粒子组成的吗？而这些亚原子粒子不又是由夸克、胶子这些亚亚原子粒子所组成的吗？到底该说包含谁才算到了尽头呢？

> “分子理论是一种可以通过代数计算出来的、非常纷繁复杂的理论，不过也许你喜欢用角度、直尺、余弦等一些熟悉的工具来求解，解了半天你自己都不相信自己到底证明了什么东西。这个时候你就得回过头再不断地检查错误、重新求解，直到相信你发现的事实与霍尔和奈特《大代数》的内容一样清楚明白，这时候再继续向前，直到你确切地全盘相信了整个事实，里面没有一丁点儿是半信半疑、像床上掉了衬衫领扣那样让你头疼的。”[2]
>
> “非常正确。”米克决定这么回答。

实际上，求解出分子是什么，是一项十分繁复的工作，这需要你从科学阶梯上较低（也许该说较深）的几层出发，一级一级向上攀爬。要想完完全全理解分子的行为特点，以及怎样通过分子来诠释物质——羊、石头、一扇窗子玻璃，解释它们为何呈现出各个方面的性质和特征，就必须懂得求解分子。不过对于很多跟分子打交道的科学家来说，大可不必跟代数扯上关系，因为这些代数大体能归结为几条分子间如何相互作用的经验法则。早在化学的数学基础完善之前，化工就已经是一个欣欣向荣的产业了。这也说明其实分子并不一定会让你头疼。

超越周期表

后来弗兰·奥布赖恩修改《达尔基档案》中福特雷警官与米克的对话，并写入著名小说《第三个警察》，这本书在他逝世后的1966年才出版。奇怪的是，他把原来的“分子理论”全部替换成了“原子理论”。这就涉及事物究竟由何组成这个颇为模棱两可的问题。到底是原子还是分子呢？化学家给出的是复合的回答，说两者皆可。他们使用的标志性暗号是元素周期表。这是个由92种天然元素（补充上一些不稳定的人造元素）通过某种化学家易于理解的模式排成的列表。“关于”化学的最著名的一本书可能是意大利化学家兼作家普里莫·莱维所写的，那本书就以这个物质基础材料的列表“周期表”而命名；它强化了人们的印象，觉得化学就该是从这个由各种符号组成的不规则表格出发。在我上中学的时候，老师就教过一种诀窍来巧记最重要的前两行元素。化学专业的本科生则要求背诵整个元素周期表，要知道铱在钴的下面，铕在钐和钆的中间。不过我很怀疑我到底有没有机会在钐身上多停留一眼（不过铕倒是会在电视机屏幕的红光中朝我们闪烁）。

元素：普里莫·莱维 的《元素周期表》[1]

在我们呼吸的空气里有所谓惰性气体。它们有奇怪的希腊名字，博学的字源，意指“新”“隐”“怠惰”“奇异”。它们真的是很迟钝，对现状极为满意。它们不参加任何化学反应，不与任何元素结合，因此几世纪都没被发现。直到1962年，一个努力的化学家，绞尽脑汁，成功地迫使“奇异”（氙气）和最强悍的氟结合。由于这功夫非常了不起，他因而得了诺贝尔奖……

钠是一种退化的金属。它的金属意义是化学方面的，不是一般语言指的金属。它既不硬也不韧；它软得像蜡，它不光不亮。除非你拼了命照顾它，不然它立刻和空气作用，使表面盖上一层丑陋的外壳；它和水反应更快，它浮在（一种会浮的金属！）水面跳舞，放出氢气……

我称了一克的糖放到白金坩埚（大宝贝）里，在火上加热。先是一阵来自焦糖的家庭气味和孩子气味，但接着火焰变成蓝灰色，气味也不一样，是金属的、无机的（事实上，是反有机的）气味——一个化学家没有嗅觉就麻烦了。至此，不太可能弄错了：过滤它，酸化它，用启普发生器产生硫化氢通过溶液。我们得到黄色的硫化砷沉淀。一句话，这里面有很毒的砷，故事中米特拉达梯[2]和包法利夫人服食的那个砷。

普里莫·莱维，《元素周期表》（1975）

4

① 以下框内译文引自牟中原译，《周期表》，山东文艺出版社2014年6月版。略有改动。——译注

② 古罗马时代小亚细亚地区的一个国王。传说他为了增强对毒药的抵抗力，每天都服食一定量的毒药。——译注

但其实化学只是偶尔才关注元素的性质，分子科学即便做不到无视大多数元素，也可以忽略其中的许多。只有在化学中极其靠近物理学的那一部分领域中，元素周期表的作用才真正显现出来，这时我们就必须拿出代数和余弦来解释为何各种元素的原子能够组成称作分子的特定结合体。周期表是19世纪一项美妙而深刻的发现，但在物理学家于20世纪创造量子力学之前，周期表也只能被当成一种神奇的密码，一种写着经验法则的小抄，提醒人们元素是分成一族一族的，同族元素有着相似的癖性。

我大概太过迅速地省略掉了元素周期表。我至少还应该再交代一下历史才对。

传统的化学史会讲，化学就是人们为理解物质进行的探索，去追问事物到底由何构成。这就将化学与古希腊哲学联系在了一起，其中就包括公元前5和前4世纪之间，留基伯和他的学生德谟克利特提出的原子构成物质的理论。化学史的这种讲述从恩培多克勒四元素说（土、气、火、水）开始，讲到柏拉图将元素理论与原子论相结合（如图1），小心翼翼地绕过中世纪炼金术士炼化物质的迷梦，谨小慎微地落到18世纪的燃素说上。然后，我们会看到1661年罗伯特·玻意耳重新定义了元素的概念（其实也未必能全然算作“重新定义”），看到老古董的四元素说在新发现的“不可分物质”前轰然崩塌；看到安托万·拉瓦锡摧毁了燃素说并用氧气取而代之，之后又于1794年在断头台上掉了脑袋。约翰·道耳顿在1800年提出了现代的原子理论，在这个世纪中元素的列表急剧扩张，接下来就由德米特里·门捷列夫将它们组织成双子座大厦形的元素周期表。铀之前空缺的部分

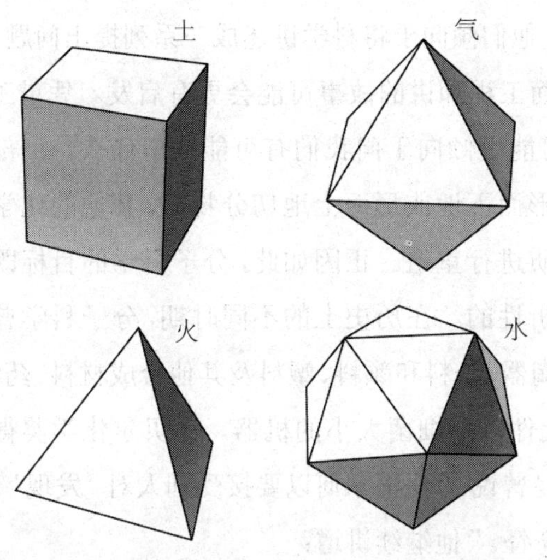

图1　柏拉图的原子。这位古希腊哲学家认为万物都是由四种元素的极小微粒构成的，这些微粒都具有正规的几何形状

又被逐步填满（铀元素本身在1789年就已经被发现了），随后沃尔夫冈·泡利及其他量子物理学家在1920年代解释了为何周期表呈现这种形状。

那么这项任务也就到此为止了。按照科学作家约翰·霍根在《科学的终结》中所写的，一旦盖上了量子力学的认证戳，就意味着化学也走到了尽头。最近又有几本关于未来科学的书暗示，化学这门学科正在从两端被侵蚀，它自己却很醒目地在学科变化中缺席了。从最基础的一侧看，化学正在朝物理学演变（包括庞大却一直被忽视的凝聚态物理这个分支，它研究的正是有形物质的行为）。从较复杂的一侧看，它又变成了生物学，生物学家正在扩大自己的地盘，研究细胞的分子机理。

6

但是，这些学术领域争夺地盘的战争，其背后的事实要有趣得多。有个奇怪的现象值得注意：很多科学史都是由物理学家撰写的，他们倾向于将科学讲述成一系列提出问题、回答问题的过程。而工程师讲的故事可能会更有启发。凭借工程师的直觉，他们可能更倾向于问我们有可能造出什么。一部分原始化学家希望形而下地或形而上地切分物质，其他的化学家则更热衷于对物质进行重组。正因如此，分子科学的目标既是创造性的又是分析性的。在历史上的不同时期，分子科学曾经分别关注过制造陶器、染料和颜料、塑料及其他合成材料、药物、防护涂层、电子元件以及细菌大小的机器。诺贝尔化学奖得主罗阿尔德·霍夫曼曾说：“化学家何以要接受别人对‘发现’的定义，这真是非常奇怪。”他继续讲道：

> 化学是关于分子及其变形的科学。有的分子本来就**在那里**，只是等着我们去认识而已……但还有更多其他的分子只是我们化学家在实验室里造出来的……[化学的]核心就是造出来的分子，无论是通过自然过程还是通过人工手段制造。

有些大学将化学系藏在了“分子科学”这样的名号之下，他们的想法可能是对的，因为这样其实是温和地将周期表这个重担丢掉，把化学家解放出来，投入到合成的世界中。这里不再是柏拉图的王国，在这里，人们设计、制造分子就是为了去**做点**什么，比如去治愈病毒感染、储存信息、固定桥梁。

普里莫·莱维是一位工业化学家，他正是工作在这个合成的世界。他对这套分子科学感到一丝歉疚：他称这种科学为“‘低级的’化学，简直像烹饪”。但“低级”化学的威力可不容小觑。它每年撬动数十亿美金的流动，它能给患病的人带来健康，能给健康的人带来疾病。汉堡和德累斯顿曾因低级的化学而化为焦土，今天在西方，人们对化学战争和生化战争的恐惧比对核武器更甚。很多人以为核弹纯粹是物理学的产物，可是写下$E=mc^2$并不能毁灭广岛，只有用铀的化合物进行同位素分离才可以。在《万有引力之虹》中，托马斯·品钦清楚地知道科学的真正威力何在：他小说中二战以后的反派并不是原子弹，而是一种新型塑料，一种称作“G型仿聚合物”的“芳族杂环高分子”，由欧洲化工巨头，包括法本①、汽巴②、嘉基③、壳牌石油④和帝国化工⑤合谋制造。这告诉我们的信息是，“实物”胜于理论。⑥

这是否意味着分子科学是坏的呢？当然不是。这只不过说明，分子科学充满了可能性：有精彩的可能，创新的可能，也有低劣的可能，噩梦的可能。有通俗而常用的事物，有奇特而少见的事物，还有人们难以理解的事物。在未来，分子科学或许能够帮助人们长出一只新的肝脏。在过去，拉斐尔、鲁本斯、雷诺阿用分子作画。而生命的本原，更是分子谱写的一曲颂歌。

① 又称染料工业利益集团，德国大型化工企业，原由多家化工公司联合而成，二战后被盟国强制解散，拆分出巴斯夫、拜耳等化工巨头。——译注

② 瑞士著名化工企业，曾与嘉基合并，后精细化工部门又独立出来。——译注

③ 原瑞士著名化工企业，现诺华的前身。——译注

④ 荷兰著名石油企业。——译注

⑤ 英国著名化工企业。——译注

⑥ 战后，盟军方面由艾森豪威尔组织起来的某个团体宣称：“若没有法本公司大量而高效的生产设施、广泛而深入的研究探索、丰富而多样的技术经验以及经济力量的高度集中，德国就不可能有能力在1939年9月发动侵略战争。”集中营使用的毒气齐克隆B正是由法本的子公司德格施生产的。——原注

合成：托马斯·品钦的《万有引力之虹》[①]

G型仿聚合物的起源可以追溯到杜邦进行的早期研究。塑料业有其悠久的传统和主流，碰巧流过了杜邦及其著名的、被人们称为“伟大的合成化学家”的工作人员卡罗瑟斯。他对大分子进行的有关研究贯穿了整个20年代，直接为我们带来了尼龙。这东西不仅使拜物教徒们欣喜万分，也为武装暴乱分子们提供了方便。同时，在圈子内部，还宣告了塑料业的一个核心信条：化学家们再也不受自然摆布了，他们现在可以决定分子有什么样的特性，然后着手制造这样的分子……这样，就可以合成出一种高分子量的单体，弯曲成杂环，拴牢，和更“天然”的苯环或芳环相串成链。这种分子链就是“芳族杂环高分子”。雅夫在二战前夕作为假想提出的一种分子链后来得到改进，成为G型仿聚合物。

托马斯·品钦，《万有引力之虹》（1973）

何为分子？

那么，一切事物都是由分子组成的吗？并不尽然。一切事物都是由原子组成的（暂不考虑某些奇怪的太空环境），而原子并非总是结合成分子。（我也说不清弗兰·奥布赖恩把“分子”

① 以下框内译文引自张文字、黄向荣译，《万有引力之虹》，译林出版社2008年8月版。——译注

改写成“原子”是因为他懂得还是不懂这其中的区别。）大多数原子本身是非常活跃的，它们天生就喜欢和别的原子结合在一起。分子就是若干原子紧密聚集在一起，分子里可以含数量多达好几百万的原子。

不过进一步来看，还有些细微的区别要加以说明。弗兰·奥布赖恩作品里的福特雷警官提到了石头“分子”和铁“分子”。但严格说来，其实并没有这种东西——至少日常的石块或者铁块里是没有这种分子的。所谓“分子”，我们一般指的是若干数得清数目的原子所聚成的分立的一团。比如在一个水分子里就有三个原子：两个氢原子和一个氧原子。一杯水里有数以万亿计的原子，但如果给这液体拍些瞬时快照（假设能看清微观细节），我们就能看到，在每个瞬间，所有原子都是三个三个聚成一团的，就像是一大群人挤在一起，可每家三个人都一直手拉手（如图2a）。

而铁的原子却并不聚成分立的分子。它们如有序的炮弹般堆积，像行列整齐的队伍。我们无法从这一大堆原子中辨认出某个小团，因为每个原子都与周围的原子间距相等。由氯原子和钠原子所组成的氯化钠晶体（即食盐，如图2b）同样如此。当铁熔化时，原子之间就会你推我搡，像混乱的人群。而当冰溶化时，氢原子和氧原子依然三个一团地手拉着手，整个晶体却分崩离析了。我们讲，冰是分子晶体——原子以分子的形式聚合在一起，而铁和岩盐则不是。

有的单质以分子的形式存在，有的不是。作为一条粗略的经验法则，铁之类的金属都不是分子，而非金属则是分子。比如，固体形式的氮就由双原子的分子组成。磷原子会四个结成

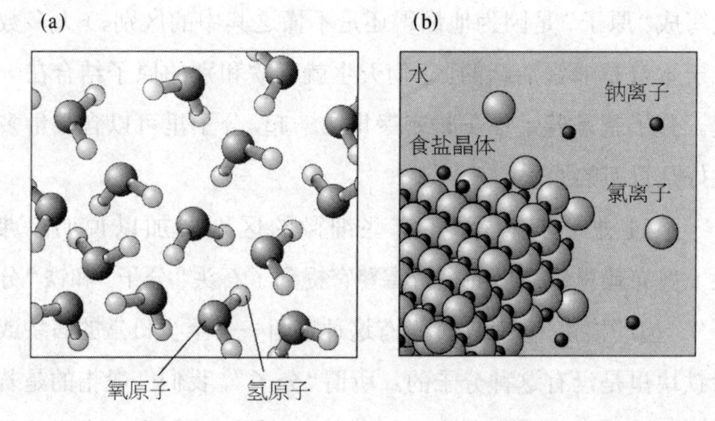

图2 水(a)由分立的三原子分子组成，三个原子通过很强的化学键结合在一起。食盐(b)则是由带电荷的钠原子(离子)与氯原子(离子)组成，其中没有分立的原子团。当食盐在水中溶解时，聚集体就会一个离子一个离子地分解开来

一团，而硫原子则会八个结成一个环状分子。我们无法只看上物质一眼就知道它的基本单元究竟是原子还是原子聚成的分子，这似乎难以接受，但没办法就是没办法。(还好科学家要找到答案并不太难。)

因此，“分子”其实是个比较灵活而松散的概念，本质上是个尺度的问题。那么，我们为什么要多此一举考虑分子，而不是直接谈论一般化的“物质”呢？我给出的理由是这样的：分子是具备化学**意义**的最小单元。在亚微观的世界里发生的故事要通过分子，而不是原子来讲述。分子是单词，而原子只不过是字母。尽管有些时候一个字母就是一个单词，但大多数单词都是若干字母按特定次序组成的分立团体。我们常常发现，比较长的单词能够传达更细致、更微妙的含义。在分子中，各个组成部分的次序也至关重要，就像“save”和“vase”含义不同一样。

分子所讲述出来的最奇妙的故事发生在生命有机体中。但很遗憾，这样的故事往往很难懂：很多单词很长、很陌生，我们对语法的掌握也很粗浅。化学家总在不断地创造出新的分子单词来扩展语言，其中有些新词还是非常巧妙的。一旦具备了某些新词，我们就能讲好以前一直不能讲的故事。还有的时候，一个新词能让我们用简单的方式来表达以前绕来绕去才能讲清的事情。

我们用语言来对分子的世界打了个比方，这个比方相当贴切。如今我们经常听到"基因的语言"，我希望让你知道，这只是分子所编码的语言中的一种。语言甚至还不只是个比喻。分子正和语言一样是切实包含着"信息"的，这一点我将在第七章中讲到。

不仅如此，语言比方的好处还在于，利用这种信息的范式来描绘分子科学具有独特的价值，它是一种会话式、应答式的描述，而不是以前那种被奉为圭臬的机械式描述。细胞生物学家越来越多地提到蛋白质分子间相互"交谈"。关注物质科学的物理学家则会讲到"合作"行为与"集体"行为。这并不是为了使科学显得更友好而刻意造出的模糊朦胧、罗曼蒂克的概念（不过若确实有这种效果也无伤大雅）。人们这样讲，正是因为他们越来越多地注意到，分子行为具有美妙的复杂性，且一般是群体性的行为，很少是单纯的线性。

考虑到这些理由，我需要在分子科学中扩大比喻的使用范围。即便是在专家之间相互交谈的层面上，我们也无法丢掉比喻。科学中很多领域都是如此，而化学尤甚。分子常常会被毫不留情地拟人化，这样做无可厚非，毕竟分子是很陌生的事物，

12

我们需要找些办法让它们变得不那么陌生。我曾经写了本关于水的书，出版商明智地坚持让我不要引入H_2O分子的球棍模型，否则就是在刁难读者，让他们把书束之高阁。可是，如果不介绍水的分子结构，我就无法解释水的奇异特点，所以我就把这些分子变成了一些小妖怪（如图3）。

我希望这样做不会有什么害处。但我最近出席一场关于分子复制的公众讲座的经历提醒了我其中的危险之处。当时听众提的第一个问题是："这些分子有意识吗？"鉴于演讲所介绍的是一种合成分子系统在模仿（非常粗糙地）某些生命有机体的

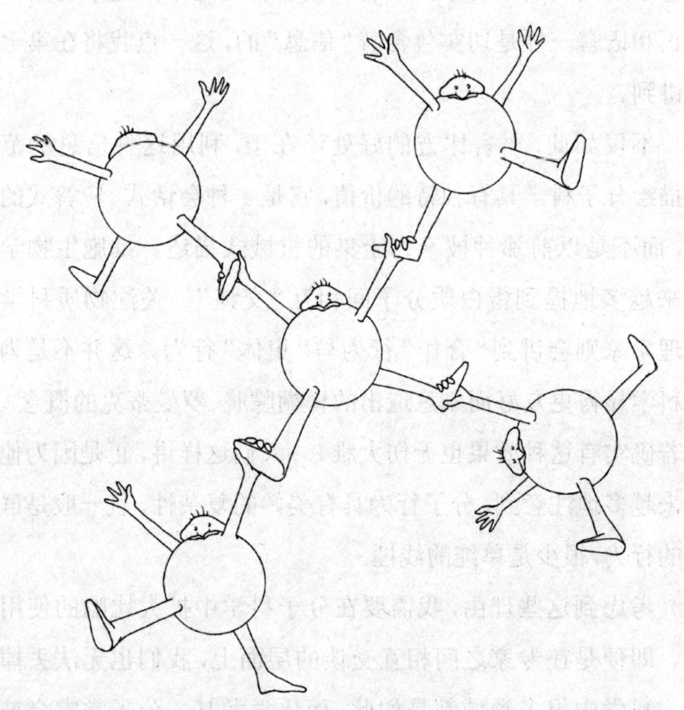

图3　分子拟人化可以帮助我们看出它们如何相互作用。我用此图说明的就是水分子之间微弱的"拉手"作用

特征，我觉得提出这个问题是可以理解的。我坚信答案是“没有”。尽管意识的概念很难把握，但基于任何对意识的切实有意义的定义，这些分子都不能算具有意识。可是，一旦我们将分子拟人化，不论结果是好是坏，我们都把意识的联系强加上去了。很多人厌恶“自私的基因”这种概念，因为它把道德判断强加其上。（理查德·道金斯称之为“诗意的科学”，我也能明白他的意思。然而，严谨的机理虽能表现成诗歌，却会被比喻的感情色彩所污染。）分子间“合作”与“交流”的想法并不能成为自然哲学的基础。不过也说不好，在分子科学中，未来或许会有某种简单、规则、有序的世界观，于是我们这些人看起来可能正像是从地心说出发解释天体运动的古代天文学家那样，强行将观察现象塞入错误的理论框架中。 13

大小和形状

普里莫·莱维的《猴子的扳手》是我所能想到的少数几部包含分子图示的小说之一（如图4）。这个分子很复杂，看上去很吓人。要是我想写本关于科学的非技术性图书，而且想给读者提供一些教益，那我做梦也不会在书里插这么一幅图的。

莱维避开了这个难题，因为他并不想让我们了解这个分子的任何事情，除了一点——该分子有形状和结构。分子里有一 14 些六边形，又有一些直线将六边形连接在一起。讲述者和一个名叫福索内的负责将大梁组装到桥上的建筑工人交谈。他说：

> 我在学校里所学习的专业也就是现在用来谋生的职业，是化学。不知道你是不是了解，不过这其实跟你的工作

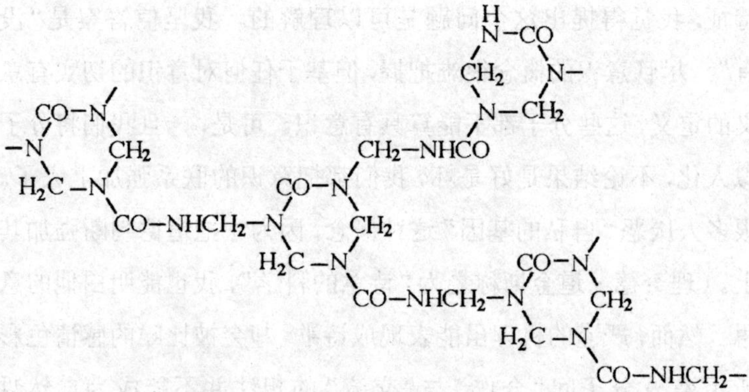

图4 普里莫·莱维的分子

有点儿像，只不过我们拆装的是非常微小的结构……我一直都是个装配化学家，也就是做合成，换句话说就是把结构排出一定的秩序。

我们在这本书中遇到的分子的例子，可以视作微型雕像、集装箱、足球、丝线、圆环、杆子、钩子，它们都是由原子聚合在一起形成的。柏拉图相信原子有“正多面体”的形状：立方体、四面体、八面体等。这是错的，[①]但化学家倒是能够把原子组成这些形状的分子。

15

那么，莱维故事中的讲述者对福索内画的这个分子到底有多大呢？图中的每个C、N之类的字母各表示一个原子，这些原子确实很微小。很多人用过很多比方来说明原子的尺度，不过

① 实际上，我们可以看到在某些例子中，柏拉图不算错得很严重。原子间的确是通过特定的几何排列方式而连接的。例如碳原子喜欢坐在由其他四个原子组成的四面体中心。这跟柏拉图所想象的“火”原子的四面体并不一样，但它能说明，柏拉图对微观世界的几何观点至少跟真理有一点沾边。——原注

我不确定这样能不能在“这种元素的不可分解的微粒真的特别特别小”以外让你留下更具体的印象。我们也打个比方：把一个高尔夫球放大成地球那么大，那么高尔夫球里的原子就像原来的高尔夫球那么大。一千万个碳原子一个挨一个地连起来，能连成一毫米长。

水分子之类的小分子，大小只有几个原子大，约为十分之三纳米（一纳米是一毫米的一百万分之一）。普里莫·莱维的分子则比这要大上几倍。（无法确切地讲究竟大几倍，因为他画出的只是分子的一个片段，它会沿图中左右两个方向不断延伸。）

分子尺度如此之小的后果之一是，分子世界里事情发生得非常快。当我们听说分子每秒能转一百亿圈时，我们大概会以为它们自转速度高得不可思议。可实际上分子实在太小了，即便以中等速度移动，也能在一瞬间就飞过分子尺度的距离。如果氧气分子要每秒转一百亿圈，那只需以每秒一米的速度运动就够了。

把原子连接在一起的小棍情形如何呢？实际上它们并不占据任何空间，而只是一种辅助我们理解图示的习惯而已。原子结合成分子时就完全挤在一起，其实相互间还会重叠，就像是两个接触的肥皂泡。这之所以可能，是因为原子并不像坚硬的台球，而更像橡皮球。它们有个坚硬且密集的中心，称为原子核。原子核大约比原子本身小一万倍，但原子的质量却主要集中在原子核上。原子核带正电荷，围绕着原子核的是一团云雾状的电子，它是带负电荷的小而轻的亚原子粒子。两个原子各自的电子云可以重叠在一起，而不致撞车，于是它们共享了一部分电子：两团云雾融合成一团，围绕着两个原子核运动。当这种情况发生时，我们就称两个原子通过**共价键**结合。上一幅分子结构图中的短线就代表共价键，这仅仅是一种辅助表示哪两个原子相互连接的办法。

谈论分子时，有一点思想至关重要，可它又不免使问题复杂化。这就是：并没有画出分子的"最佳"方式。有人可能会说：不要管结构图了，为什么不直接画出它们"真实"的样子呢？但这办不到，因为我们无法像给猫或树照相那样去给分子照相。这不是技术水平局限的问题，不是因为我们缺一台能分辨如此微小物体的显微镜或照相机，而是因为"看"的机制本身就不允许我们"看"一个分子（或一个原子）"本来的面目"。

原因是我们只能够看到可见光，可见光是波状的辐射，它的波长（相邻两个波峰间的距离）范围是从700纳米左右的红色光到400纳米左右的紫色光。换句话说，一厘米内包含有约14万个红光的波形。这样的波长是分子大小的好几百倍。大致说来，光不可能聚焦到一个小于其波长的点上，也就是说那么小的物体是无法被可见光分辨的。[①]因此，基于可见光的显微镜是不可能给出水分子的清晰图像的。

我怀疑这可能是人们认为分子难以理解的原因之一，也是为什么像前面那样的结构图能让科学图书吓跑读者的原因。这种东西不仅小得看不见，而且小得都不可以"看"了，还要具体地谈论它岂不荒唐？看不见的东西就有种迷幻的气质，好像只

① 我这里谈的是传统的显微镜，即利用透镜进行聚光。现在有一些新的光学显微镜能够突破波长极限分辨率，其原理是使光源非常接近样品，并从微小的孔径中射出。迄今为止，这种方法能够将分辨极限减小到波长的十分之一。——原注

是杜撰而已。

不过分子可不是杜撰，我们不仅能够证明它们存在，还能证明它们有确切的形状和大小。图5给出了一些分子的“肖像”，是通过一种非光学显微镜所成的像。每幅图边上我都附上了分子结构图。早在这种显微镜发明之前，人们就已经知道这些分子是这样的结构了，但从没有人能够直接看到它们。这些图像挺模糊的，单从这些图像入手，你没法猜出分子的准确形状。但显微镜下显示的形状与我们所预期的完全一致，非常令人信服。

在照片拍摄以前，我们又是怎么知道这些分子的形状的呢？从实验中能够得到一些确凿的证据。尽管分子实在太小，无法被可见光分辨，我们仍然可以通过波长与分子大小相当的辐射来“看到”它们。波长约十分之一纳米的辐射属于X光，通过让X光在晶体表面反射，就有可能推断出构成它们的原子处在什么位置上。也就是说，如果物质可以制成结晶态，使分子有规律地堆积在一起，那么使用这种名为“X射线晶体学”的技术手段就可以揭示分子的结构。

原则上我们可以用X光看见单个的分子，只要像光学显微镜汇聚可见光那样将X光汇聚起来就可以了。但实践中要聚焦X光十分困难，现在仍无法做到，不过科学家们几乎快要实现它了。同时，我们还可以使用电子显微镜，即将一束电子打到样品上反射，并进行汇聚，得到图像。电子也可以表现得像波一样。利用电子波，蛋白质或者DNA（脱氧核糖核酸）等大分子的图像我们也能构建出来。这些图像的细节不够充分，不足以显示出单个的原子，但能让我们对分子整体的形状留下印象。

18

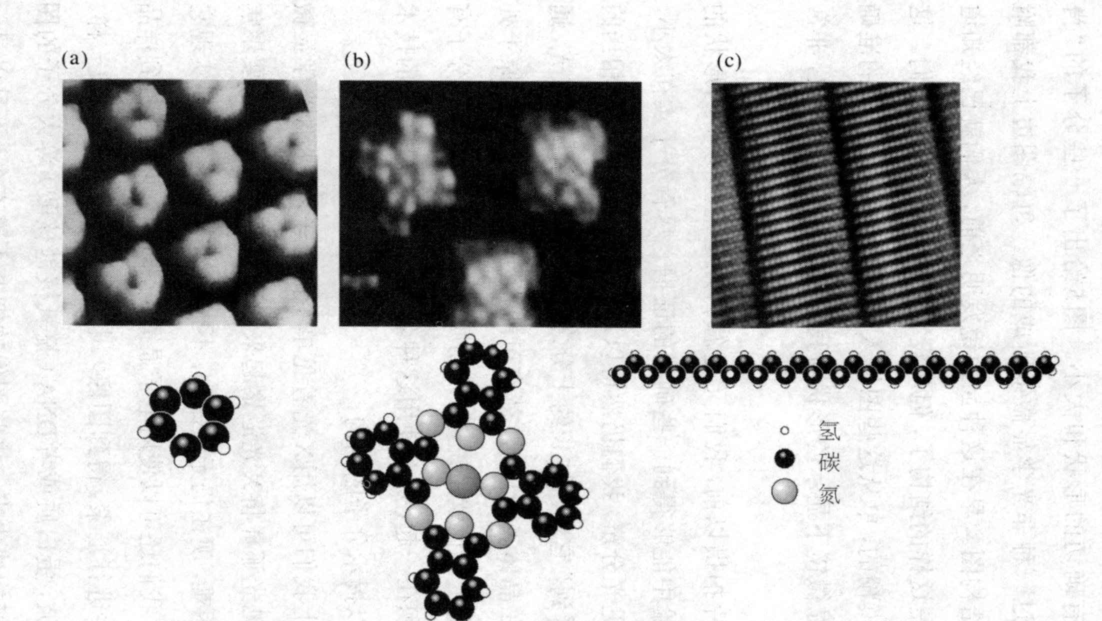

图5　扫描隧道显微镜“拍摄”下的分子。扫描隧道显微镜可以逐个分辨分子，但还不能够显示出足够（目前为止？）的细节。若事先缺少了解，仍然无法对图像进行解读

另一种推断分子形状的方法则是理论的方法：我们是有可能计算出它们的。这就要涉及弗兰·奥布赖恩讲“分子理论”提到的“代数”，不过在此没必要细讲。只需要了解，量子力学[①]的定律能够使我们预测原子间如何成键，及原子相互位置如何。原子是不能随心所欲地结合的。比如，各种元素的原子都倾向于形成固定数量的化学键，这个数就称为它的化合价。碳原子喜欢成四个键，氢原子喜欢成一个键，氧原子则成两个键。

分子结构的量子理论确实是一种“非常纷繁复杂的理论”，即使用最好的计算机也只能近似地求解方程。但现在，我们对中等大小分子结构的求解已经能够达到相当的可信度。计算预测结果与X射线晶体学测定的分子结构进行比较，常常高度吻合。不过要预测生物细胞中发现的诸多大分子的形状，我们还没有可靠的办法。这种情况下，X射线晶体学也很困难，原因有二：一是因为这些分子晶体散射出来的X射线图样很难解读，二是因为很多时候这种分子无法形成晶体。所以细胞中充满了我们不知道其形状的分子。

分子的形状正是它如何行为的关键因素，所以我们要理解生命分子如何工作就遇到了很大的障碍。把一位设计师的格言倒过来讲就是——“功能服从于形式”。

总之，分子科学是一种高度可视化的科学。化学家花了两百多年发展出一套描述这些分子的图形化语言，结果他们现在必须讲多种语言。描绘分子有多种方法，不同方法各有其侧重点，着重表现描述者想强调的方面。英国化学家约翰·道耳顿

20

① 量子力学是对极小尺度下（通常是原子尺度）物质及其行为的一种数学描述。在这样的尺度下，物质会表现出波的性质。——原注

从1800年开始把分子画成原子的集合，而用圆圈符号表示原子，每个圆圈有阴影或标记，用来区分不同的元素。一旦知道了对应关系，这种表示法就很清楚了（如图6）。

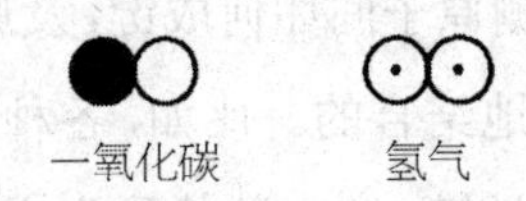

图6　道耳顿的分子

这方法不错，不过对印刷工来讲可不轻松，他们得专门补充上这些符号。一种更简洁的办法是用一到两个字母符号来表示不同元素：C表示碳（carbon），O表示氧（oxygen），Ca表示钙（calcium），Fe表示铁（iron）。（甚至到19世纪时，化学家仍然将这种金属记作拉丁文的*ferrum*。同样的原因，金和银分别称作*aurum*和*argentum*，于是用Au和Ag表示。Ir不表示铁元素而表示铱元素。不过至少创制之时人们是**希望**这套体系不言自明的。）

于是，一氧化碳就可以简记为CO。相同元素的多个原子可以用下标表示，于是氢气分子就是H_2。

但这套方案并没有给各个分子以独一无二的表示。甲醚和乙醇是不同物质，性质不同，但它们的**化学式**都表示为C_2H_6O。我们又回到了之前的那个词汇学问题：一个词语的意义不仅由它包含什么字母决定，也由字母的排列顺序决定。

所以我们还需要某种新形式，它能够表示出原子间如何相互连接。这时候字母间的短线就来了，它表示化学键。C_2H_6O的两个版本——称为两种**同分异构体**（组成原子相同，排列顺序不同）——可以表示成如下的样子：

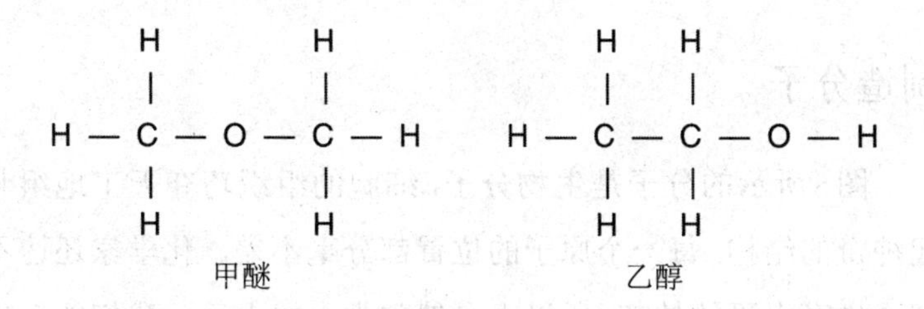

更加复杂的情况是，分子不是纸面一般的二维形式，而是会占用整个三维的空间。

表示第三个维度有几种不同方法，有了计算机图形学的出现，我们的方法更显精巧了。图7所示是对中等大小分子的一种立体表示法：将两只眼的图像重叠，就可以看到3D形状。

化学家设想出来的办法远不止这些。有时我们还需要“空间填充模型”来表示分子占据了多大的空间（如图8a）。有时则大体的模式图最为有用，这时就不用表示出不必要的细节（如图8b）。

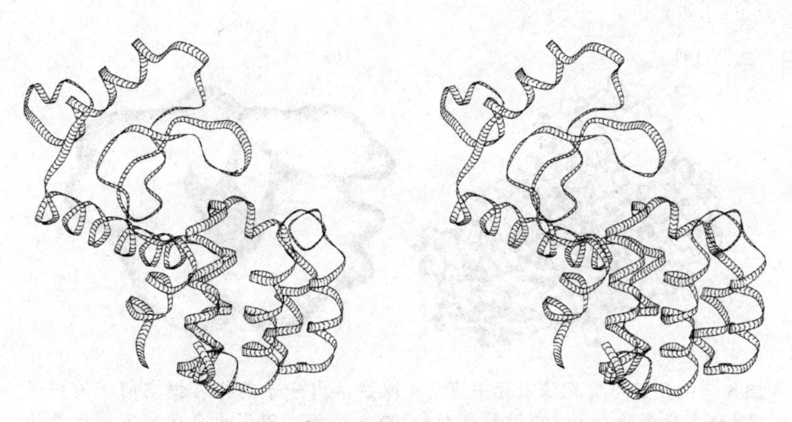

图7　这两幅立体图可以使我们看出三维空间中的分子形状。这里表示的分子是唾液中存在的一种溶菌酶。图中可以清晰地看到蛋白质链的螺旋部分。眼睛距此页20厘米，双眼视线向中间相交，这时可以看到三幅图。集中注意中间的那幅，多看几秒，它就会变得清晰起来

制造分子

图8所示的分子是生物分子，细胞的组织巧夺天工地组装起神奇的结构，每一个原子的位置都分毫不差。化学家还达不到这样高水平的技艺，所以大自然常常占据上风。我们能造出杀死致病菌的分子、消灭病毒的分子以及摧毁癌细胞的分子，但它们开展工作常常很粗暴。它们常能发挥惊人的功效，但同时也可能会在作用过程中破坏健康的细胞；要不然就是入侵的有机体很快找到办法魔高一丈，比如细菌逐渐演变出对抗生素的免疫力。不过化学家制造分子的手艺越来越精湛，进展迅猛，也许有一天，造出保证有效且全无副作用的药品就不再是梦了。

普里莫·莱维这样写道：

> 认真说来，其实我们是很糟糕的装配工。我们就像笨

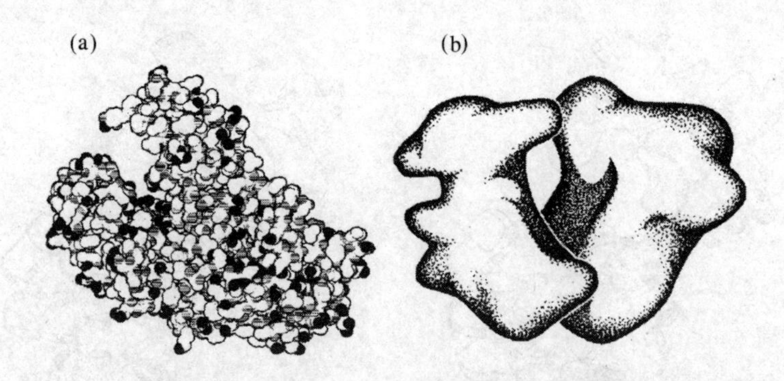

图8　(a) 空间填充法表示分子，可以显示出分子怎样占据空间。该分子是DNA聚合酶分子，能够制造新的DNA分子。阴影用来区分不同种类的原子。(b) 若对原子尺度下的结构认识不完备，抑或只是想避免太多细节，有时就需要大致的模式图。这里画出的是核糖体复合物，能够制造新的蛋白质

拙的大象，别人给了一个封闭的盒子，里面装着一只手表的所有零件。我们很健壮，很有耐心，于是就拿着盒子用尽全力朝各种方向使劲摇啊摇。我们也可能会给盒子加加热，因为加热也算是另一种形式的摇晃。呵，有时如果手表不算太复杂，只要我们不停地摇，那迟早能把手表组装到一起……

这其实是最糟糕的情景。莱维是在1978年写的这段话，自那以后情况有了长足的发展。可是直到20世纪后一二十年，莱维所描述的方法——化学家愿意称之为“颠热锅”——时常也只是他们能用的最好办法。合成化学的主要工作就是专注于制造所谓的“有机分子”，这意味着分子包含主要由碳原子构成的骨架。到现在为止我所描述的大多数分子都是人们认为的有机分子。在甲醚和乙醇分子中，碳骨架非常小。但对于普里莫·莱维的分子而言（第15页），它的骨架就比较复杂了。你还能看到，氮原子在它的骨架中也占有一席之地，而甲醚分子中的氧原子像一座重要的桥梁。有机分子并不一定只由碳原子构成骨架，不过是碳占据主导地位罢了。

“有机”这个词似乎选得有点奇怪，因为有机化学家摆弄的分子大多数并非自然界有机体的产物，而是源于实验室。用这个词有历史原因，因为有机化学一度就是研究生命有机体中的分子的学问。现在我们知道，这些分子主要是基于碳的。为什么是碳呢？这是因为碳在各种元素中具有最特殊的能力，能够以种种复杂形状连接成稳定的分子框架，包括环形、长链、分支网络等。

24

19世纪的化学家对制造新的有机分子充其量只有点模糊的意识。他们可以对来自自然界的分子进行修改，切掉一小段碳骨架，用另一段来代替。但要对整体的框架本身进行改变则比较困难。特别是，他们经常对要制造的分子的真正结构一无所知，这就更是雪上加霜了。他们仅凭着“颠热锅”的办法就造出了第一例合成塑料、合成染料和合成药物，这真是奇迹。

从当时某些化学家所处的情景下出发，我们就能看清，分子建造在做些什么，何以会如此困难又如此让人着迷。在1850年代中期，在伦敦工作的德国化学家奥古斯特·威廉·霍夫曼，指导其十来岁的学生威廉·珀金用已提纯的煤焦油组分制取奎宁。奎宁是金鸡纳树的一种天然提取物，可以治疗疟疾。煤焦油则是煤气厂里大量的黑色黏稠残留物，自19世纪早期发明了煤气灯以来就迅速涌现。煤焦油不是一种很有前景的原料，但霍夫曼等人还是发现，通过蒸馏可以从中分离出若干种富碳的芳香族有机化合物，如苯、甲苯、二甲苯和苯酚。

没有人知道这些化合物的结构，没有人能够画出我们之前所展示的那种连接原子的短线图。时人只知道化合物中含有各种元素各多少，也就是知道化合物的化学式。比如苯的化学式是C_6H_6，奎宁的化学式是$C_{20}H_{24}N_2O_2$。而分子中的碳骨架是什么形状则一无所知。

霍夫曼的（也就是珀金的）办法是数原子。他们从煤焦油的一种提取物出发，转化得到一种叫作“烯丙基甲苯胺”的化合物。这种化合物与奎宁所包含的元素基本相同，元素的比例也大体一致，于是他们就希望对这种化合物进行某些适当的操作，从而将它转化成奎宁。他们猜测，两个烯丙基甲苯胺分子（化学式$C_{10}H_{13}N$）再结合上一些氧和氢就能造出这种药物。但希望其实很渺茫，因为十个碳原子结合的方式多种多样。而事实上，丙烯基甲苯胺的碳骨架跟半个奎宁分子也的确并不一样。

因此，珀金在伦敦东部父母家里临时搭建的实验室里的实验并不成功，结果只得到了些铁锈色的污泥——有机化学家并不喜欢见到的老朋友。可是年轻的珀金没有放弃，他换掉了丙烯基甲苯胺，从另一种叫作“苯胺”的有机化合物开始。这回，污泥成了黑色。用甲基化酒精去溶解，则呈现出壮丽的紫色。珀金兴奋地发现，这种紫色可以用来染丝。于是他发现了第一例苯胺染料。珀金和父亲、哥哥一起开设工厂制造这种染料，很快在英国和法国开始大规模投产。这不仅标志着合成染料工业的开端，也是整个现代化学工业的肇始。许多当今的化学企业，如巴斯夫、汽巴-嘉基、赫斯特等，都是从生产苯胺染料起家的。

到了19世纪后25年，有机分子合成的随意性已经降低了。奥古斯特·弗雷德里克·冯·克库勒在1857年推断出碳是四价元素，即倾向于形成四个化学键。在1865年他提出，与所有芳香族煤焦油分子有关联的苯，包含着由六个碳原子组成的**环**，这个结构成了有机化学中无所不在的主旋律。1868年，德国化学家卡尔·格雷贝与卡尔·利伯曼合成了茜素分子，这种物质正是用茜草根提取的染料中红色的来源。这种红染料是商业上几种最为重要的自然染料之一，而有了格雷贝和利伯曼的合成法，人们就可以更廉价地得到这种人工产品。

茜素的合成是分子制造历史上的里程碑，其原因有二。首先，人们从初始材料（另一种煤焦油中的芳香成分，称作蒽）出

26

发，经过设计好的修改步骤，最终得到了产物，而不是拿着原材料“乱炖”并祈求好运。化学家们不仅知道蒽的化学式，还了解蒽的化学结构，知道它的结构正与茜素相关。（其实他们把结构猜错了，不过幸运的是最终结果没受到影响。）有机化学家把这样的过程称作**定向合成**，即将起始分子系统地一步一步转化，最后得到目标产物。

其次，格雷贝和利伯曼在实验室中造出了茜素，这显示了有机化学的威力足以匹敌大自然。人们有可能造出从生命有机体中找到的复杂分子——化学家今天称之为**天然产物**。

那么，格雷贝和利伯曼合成的、后来在化工厂中成吨制造的合成红染料与天然的茜草红染料是完全一样的吗？既是又不是。从茜草根中用传统方式提取的染料其实是几种不同化合物的混合。茜素是主要的呈现颜色的分子，但提取物中还包含一种非常相关的组分，称作红紫素，呈现橙色（和名字有点不一样）。而将蒽转化为合成茜素的过程中也会产生几种副产物，主要是一些与茜素结构十分相似的分子。珀金和其他一些化学家在1870年代前期辨别出了茜素工业生产过程中至少四种副产物。毫无疑问，还有其他很多种副产物以更少量的形式存在。

所以，合成的茜素分子与茜草根中提取出的茜素分子是相同的，但实际得到的合成染料与天然的染料则有所区别——它们都不纯净。对任何“天然产物”进行工业化合成制造，都会有这样的情况，因为有机化学家使用的一切合成过程都会产生副产品。不过这并不能说明合成的化学物质就比从自然中提取的等价物更好或者更糟，毕竟它们都是某种程度上的非纯净物。

但化学家对纯净度有很高的追求，会花大量时间从产物中去除杂质。而天然提取物则是比较复杂的混合物，除非经过加工分离了各种成分。

煤焦油衍生物质的商业价值还不仅仅是用作染料而已。德国医学家保罗·埃利克在1870年代利用新合成的染料对细胞进行染色，以便于在显微镜下研究。他注意到，细菌细胞吸收了某些染料之后会被杀死，这提示它有可能应用于医疗。埃利克开始合成一些染料化合物，把它们当作药物来测试。1909年，他用这种办法找到一种能杀死引起梅毒的寄生生物的含砷染料。自中世纪以来人们一直使用汞，而这种名为“洒尔佛散”的药物第一次缓解了这种致死的疾病。这正是现代化学疗法之滥觞。

19年后，亚历山大·弗莱明发现了青霉素。这是一种霉菌产生的化合物，能够杀灭细菌。这是首例抗生素。它大大降低了伤口感染的风险，给外科医学带来了一场革命。还有很多其他的自然产物，也在生理学上发挥了有益的作用。比如水杨酸，这是柳树皮中的一种提取物，既能抗菌又能止痛。和它紧密相关的另一种分子就是阿司匹林的成分，是由拜耳公司在1899年制造的。化学家与药学家不断在自然界的分子武器库中搜寻潜在的药物，而后再找办法来合成那些有效的。

这其中有一种化合物，近年来声名鹊起，名字叫紫杉醇。这是太平洋紫杉树中的一种天然产物，1980年代人们发现它在阻碍细胞分裂方面非常有效。而癌症正是细胞失控增殖的结果，于是紫杉醇就成为有潜力抗击癌症的药物。美国食品药品管理局已经批准它用于治疗乳腺癌、肺癌、卵巢癌以及前列腺癌。但紫杉树并不是一种可靠的药物来源，从每棵树中只能提取出几

28

毫克的这种物质，而且只能从树皮中提取，所以一旦取走紫杉醇，树木就死亡了。紫杉树本来已是濒危物种，就算全部灭绝也无法满足全球对紫杉醇的需求量。显然我们需要合成它。

紫杉醇的分子结构相当复杂。它的骨架包含了四个碳环：一个四原子环、两个六原子环和一个八原子环（如图9）。不同的附属原子团挂在这个骨架上。当前没有任何标准化学试剂具有这种骨架形式，我们必须从头开始搭建。

这对化学装配工来说是一个深刻改变。普里莫·莱维笔下的人物说明了合成有机化学家今天又是如何工作的：

> 你能想象到，其实更合理的办法是逐步向前推进，先将两块连接起来，再加上第三块，等等。这（比"颠热锅"）需要更多的耐心，但的确能领先一步。大多数时间我们都是这么做的。

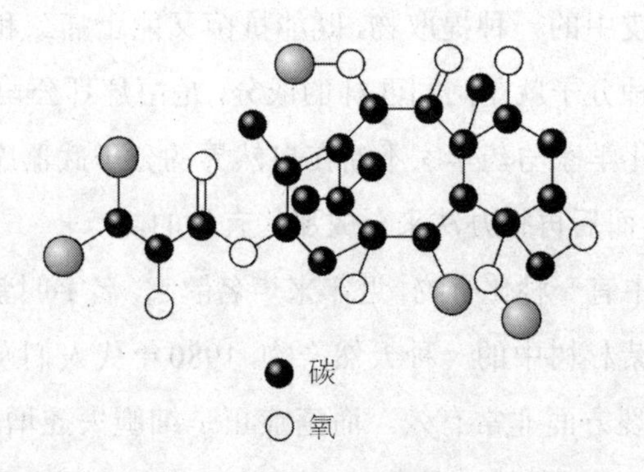

图9　紫杉醇分子，图中只画出了骨架中的关键元素：黑球表示碳原子，白球表示氧原子。大的灰球表示"取代基"原子团，里面包含碳原子、氧原子和氢原子，在此省略表示。在分子中还有其他一些氢原子，我也把它们略去了，以保持画面整洁

类似这样的合成需要事先精心规划。最常用的规划法是美国的诺贝尔奖得主伊莱亚斯·J. 科里所设想的。他把这种办法称作“逆向合成分析”。顾名思义，就是从目标产物出发，回退分析，就像拆卸一个分子模型那样。每一步，你都选择拆掉一个你能看出怎样再装上的键，这样一来，在向前推进的过程中你就知道怎样进行每一步连接了。问题的诀窍在于要回退到起始原料，即碳骨架的一些片段，它们要么现成可用，要么就是能简单地用现成的化合物来合成。 29

在紫杉醇的例子中，有两个课题组“领先一步”。一个课题组在加利福尼亚的斯克里普斯研究所，由K. C. 尼古拉奥领导；另一个课题组由罗伯特·霍尔顿领导，在佛罗里达州立大学。1994年，他们在相差不到一周时间内都给出了多步合成过程。要制造这样一个复杂的分子，并非只有唯一的办法，甚至也没有“最佳”的办法——研究者们在那之后又报告了几种其他的方案。但对于大规模生产来说，所有这些方案都太过复杂，无法实施。所以紫杉醇目前是用“半合成”的办法制造的，即先从紫杉的针状树叶中得到一种中间化合物，这种物质很像紫杉醇的半成品。使用这种物质，在实验室中就可以比较高效地完成合成，而且摘取树叶也不会杀死树木。 30

我在这里谈的都是有机分子的制造，但我还是应该强调，也有很多化学家并不基于碳原子来建造分子，而基于其他种类的元素。这些分子通常都很小，因为其他元素不能很稳定地形成像碳那样很大又很复杂的骨架。不过这条规则也有反例，图10中的分子就引人注目，这是一个主要由钼原子和氧原子组成的环。它是由德国比勒费尔德大学的阿希姆·缪勒课题组制成

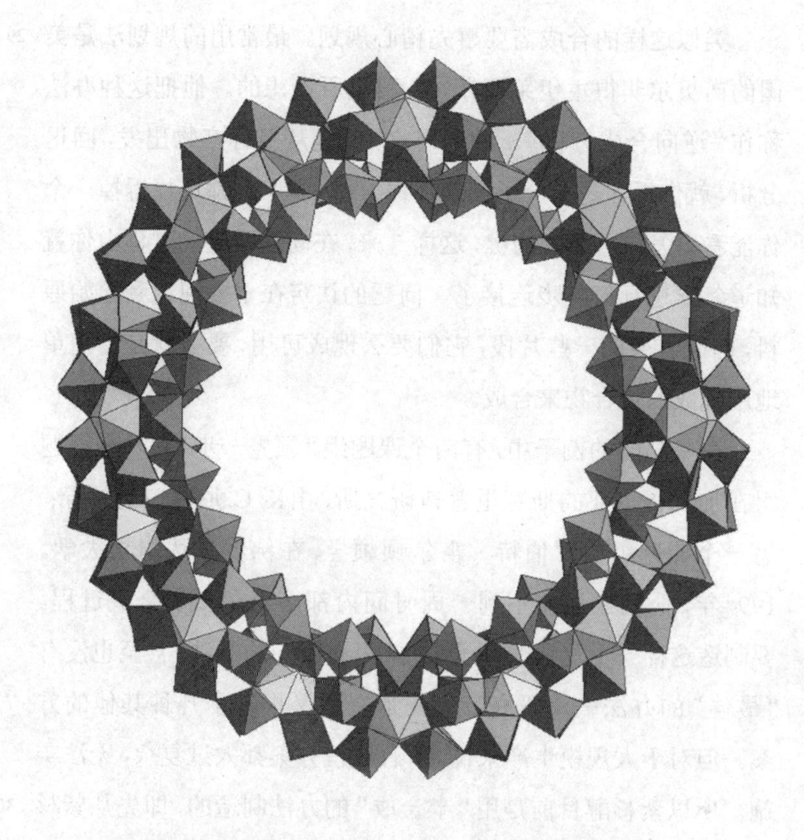

图10　德国阿希姆·缪勒课题组制成的分子“大轮胎”(顶视和侧视)。每个金字塔形都是一簇钼原子和氧原子

的，大小足有4纳米（水分子宽度的15倍，人类发丝宽度的数万分之一）。当金属与氧相结合时，它们一般并不会形成大分子，而是要么形成一些只包含寥寥几个原子的小分子，要么结晶形成矿石一样的固体（你要乐意也可以叫它“无穷大分子”）。化学家近来对这种**无机**大分子很感兴趣，因为它们能够表现出不一般的、可能会有应用价值的特征，如磁性或者导电性。芯片里的晶体管之类的电子元件就是由无机材料制成的，主要材料是硅和硅的氧化物。当今复杂的分子烹饪术给我们列出了花样繁多的菜单，为分子尺度的电子科技定制元件也只是其中的小小一项。

第二章

生命的征象：生物分子

科学家与诗人达成一致令人略感欣慰。1949年，英国生物学家J. B. S.霍尔丹因“什么是生命”这个问题陷入了深深的思考。在文章开头处，他这样坦陈：

> 我并不打算回答这个问题。其实，我甚至怀疑究竟有没有可能对这个问题给出个完满的回答。因为我们都知道活着的感觉是什么，就像我们能感觉到红色、疼痛和力气。我们却无法用别的什么词语描述它们。

艾米莉·狄金森说得更加简练：

> 自然真意人皆悟，道常无名未可言。

而霍尔丹则继续探索道：

> 生命是种化学过程的模式。这种模式有特殊的性质。它和火焰燃烧的模式相似，但又比火焰多了自我调节的功能……因此，当我们说生命是一种化学过程的模式时，我们说出的内容确切而且重要……但若要全然照此解释生命，则无异于把生命还原成机械作用，我相信这是不可能的。[33]

生命是否只是些相互作用的分子，极其繁复却也能从原理上阐释？抑或生命还涉及更多的东西？我们还不知道。科学家采用自底向上的方法来研究，他们做出尽可能少的假设，只提出能够加以检验的学说。这种方法会不会最终触及一个极限，那之后科学再也无力延伸，这我们说不清。不过，现今这样的终点还没有明显出现。生命——我们可以宽泛地将其定义为能够繁殖、能够对外部条件做出反应并从环境中获取养分的有机体——确实有可能仅仅只是分子及分子间的关系而已。而且，这种可能性确实极大。不过这并不令人失望，反而是非同凡响的。分子们串通一气竟然能创造出《李尔王》，正是这种无限的可能造就了这个神奇的世界。

可是，我也认为人类的思想（更不用说思想所创造的奇迹）永远不可能仅用分子的理论来**解读**，就像我们也不可能用字母表来解读**李尔王**。大多数科学家也都相信如此。现象是分层级的，我们不可能只考虑发生在第一梯级上的事情，然后就理解所有梯级上的事情。无论我对晶体管工作原理了解得多么透彻，我都不可能从中推断出为什么电脑会死机。如果我播下种子没有发芽，那最好应该考虑土壤中养料、湿度和温度的问题，而不该对种子进行基因分析。从事科研的很多技能都有赖于弄清你在研究哪一个层级，进一步讲就是要明白什么与研究相关，什么无关。

在我们探索生命世界的分子之前把这些说清楚很有必要，因为生物学的分子观点常常被贴上还原论的标签，即试图在基因的分子层面上解释生命的方方面面。有时候，分子观点确实是推进研究的最佳方法，因为分子毕竟是生命基础的最小功能单元。但如果我们像多数科学家一样能够同意，在阶梯上降到微观世界层级就必须抛弃其他范围的关于生命的问题和答案（比如：意识是什么？），这样的下降就不会引起明显的异议了。

实际上，通过分子观点这条途径，我们还能更清楚地认识我们自身的自然基础。分子生物学帮助人们填补上了查尔斯·达尔文进化论的主要鸿沟，即自然选择的**机制**问题。分子生物学至少让我们略知一二，生命如何出现在这遍布气体、岩石和水的星球上。它拯救了生命，缓解了痛苦。它帮助我们理解，为什么药物并非总是如愿发挥效用，为什么抗生素的滥用会培养出超级病菌，艾滋病毒怎样为非作歹。对生命分子的研究成为20世纪后半叶的主流科学，而且看起来还将在我们未来的生活中发挥更显要的作用。这个科学领域中的知识可能将会越来越普及，而不再是奢侈品。

生命活力

人们一度认为，有机化学和别的化学不一样。19世纪早期，不少科学家相信有机物质是生命有机体中生命活力发生作用的产物，化学家是不可能在实验室中仿制的。但到了1818年，颇有名望的瑞典化学家约恩斯·雅各布·贝采里乌斯从生命活力的思想中隐约看到了某种循环证明，阻挡了人们进一步发展的希望：

> 从我们的观点来看，动物体内多数现象的原因都隐藏得很深，永远无法找到。我们将这些潜藏的原因称作**生命活力**。和很多其他的概念一样，前人欺骗性地导向这样的论点只是徒劳，而我们为自己则创造出这个**词语**，不能附加新的观念。

35

紧接着，贝采里乌斯又提示我们怎样更上一层楼：

> 这种**生命的力量**，不是我们身体的组成部分，也不是身体的某种机能，也不是一种简单的力量。实际上，它是身体机能与原料之间相互作用的结果……

这正是关键。理解生命的分子基础，与其说是要去搞懂生命分子都是些什么，不如说是要理解生命分子相互间做些什么。生命的分子本性并不是一场陈列，而是一支舞蹈。在后面的章节中，我会叙述其中的一些舞步。而在这里，我想简要地介绍部分角色。

在霍尔丹的时代，人们常将生命看作一系列化学变化，就像是把实验室里的玻璃仪器连成巨大的网络。科学家们相信，一切的关键就在于代谢作用，即我们如何从食物中获取能量。但是，光把细胞的组分分子提纯扔到锅里可造不出有机体来。现代的分子生物学则关注时间与空间下的**组织**。生命分子在细胞各个隔间中是如何排布的？它们怎样在周围移动？它们怎样相互沟通来统一步调？我们现在之所以能够提出这些问题，正是因为我们已经可以在分子层面上来研究细胞，对工作的分子进行测量和拍照。于是，细胞成为了一个社群。

不过这个社群可是相当复杂的。分子生物学的困难之处与理论物理学并不相同。它的概念并不陌生、抽象，或者在数学上很艰深。分子生物学之所以困难，是因为同时发生的事情太多。当我们的分子机器运转错误时，我们会有惊吓甚至休克的反应，但更惊人的是它竟然还是完全可以运转。分子机器经常如此，36 它设计的目标就是要十分牢靠，以面对这个世界的变迁。分子机器里有检查点，有安全机制，有备用预案，还有细致的记录存档。没有任何人造机器能达到细胞这般缜密的组织程度。

我们应当记住，细胞是个自动机器的社群。社群的各个成员没有意志，没有远见，没有记忆，没有利他之心（严格地讲也没有利己之意）。可是它们常常优美地共同协作，导致我们很容易忘记上面这一点。另一方面来看，细胞也会捉摸不定，因为我们对它们怎样工作了解得太少了。当我们预期它们死亡时，它们却可能存活下去；当我们给它们提供试验的药物时，它们可能发生出乎意料的反应。

分子生物学在细胞层面发挥效用，而很少谈论整个生物体。细胞可谓是“生命的原子”——你无法找到更小的存活单元。（病毒是一种具有争议的例外情形，它们只不过是基因穿上了外套，但除非去感染细胞并攻击其组织，否则病毒是无法繁殖的。）但这也并不见得说明这种视角很受局限，因为我们可以根据单个细胞的内部活动，来理解相当多的我们人类的需要。人体细胞需要氧气和糖类来制造新分子并且自我复制，所以我们需要呼吸和进食。神经刺激起始于细胞的层面。人体各种组织——皮肤、毛发、骨骼、肌肉——都是细胞里的分子一个个堆积起来的。我们排泄是为了将细胞中的垃圾清走。发抖、出汗都是我们为了稳定细胞温度的措施。换言之，很多关于人体功能的问题都可以在分子生物学的层级上得到回答。当然了，也有很多不能这样解答的问题，它们多是最有意思的生物学问题。

细胞里的大腕

霍尔丹将生命是“蛋白质的存在方式”这一命题的提出归于恩格斯。（霍尔丹是一位社会主义者，并非为了生物学而阅读恩格斯的著作。）这种活力论观点隐含了蛋白质内在具有生命之意，但霍尔丹抨击了这种想法。不过霍尔丹并不反对的观点是，**蛋白质**乃是生命的材料。

蛋白质在生命细胞中处处皆是。很多蛋白质是酶，即催化化学变化过程的分子。酶能够以数百万倍的系数加速化学反应，确保体内的化学反应不致慢得难以想象。酶是在研究发酵的过程中被发现的，酶的英文enzyme正是希腊语“在酵母里”的意思。19世纪晚期，人们发现从酵母细胞中可以获取酶并提纯，虽然酶已经不再是生命系统中的一部分，但它还是可以继续进行发酵。这项发现帮助人们认识到，生命中的化学同样也遵循与非生命物质相同的原理。

如果细胞是座城市，酶就是其中的工人。为了维持城市运转，原始材料输入细胞并转化为有用的东西。酶就是工厂里的工人，推动这项工作完成。而这个产业里令人好奇的一点是，它还包括负责制造工人本身的工厂——酶自身也是在生产线上组装起来的。

不是所有的蛋白质都是酶。有的蛋白质发挥着结构方面的作用，成为身体的各种组织。有的作为细胞的警卫力量。有的在蛋白质的运输轨道上负责来回递送包裹。有的负责操作细胞的大门，坐在细胞外膜上，遵照接收到的指示开门关门。在人体细胞中有约六万种不同的蛋白质分子，每种都负责一项高度专门化的任务。

如果只是瞧一眼蛋白质，几乎不可能猜出它负责的任务是什么。从外观上来看，蛋白质没什么特殊之处，大多呈球形（参见第23页图8），主要由碳、氢、氧、氮和少量的硫组成。负责同一项任务的蛋白质具有相同的形状和结构，那些看似不规则的一团团实际上是精心设计并装配好的。

很多酶的形状像凹凸不平的腰果，在内侧的曲线处有裂缝。这个裂缝就是要执行任务，即分子实施催化的位置。有些蛋白质会成组地完成工作，它们变成了多蛋白集合体中的“子单元”。细菌体内的色氨酸合成酶就是如此，由四个可分离的子单元结合而成。这种酶能够合成色氨酸这种小分子。所有生物体都需要色氨酸，但人体不含这种酶，所以就不得不通过吃掉其他已经造出色氨酸的生物来摄取它。

这个例子展示了，酶及其他蛋白质的名字常常能透露它们的功能。乙醇脱氢酶是能从乙醇分子上取下一个氢原子（“脱氢”）的酶。ATP（三磷酸腺苷）合成酶能合成ATP分子。但并非所有蛋白质的名字都如此一目了然。血红蛋白（haemoglobin）负责在血流中携带氧，它的名字则分为两段，前段来自希腊语的血液（*haeme*），后段来自它的形状是球形（globular）。肌红蛋白（myoglobin）在肌肉组织中负责从血红蛋白手中接过货物——氧，它名字的前一段来自肌肉的希腊语。还有更诡异的名字。弹性蛋白是一种有弹力的蛋白，能在很多柔韧的人体组织，如血管和声带中找到。泛素是人体中到处都能找到的一种蛋白质，因为它在随处可见的废弃蛋白质处置过程中发挥核心作用。

从图8中你大概猜不到，其实蛋白质分子只不过是小分子串起来的一条单链。这条链将自己密集地折叠、盘绕起来，看上去就像是堆起来的一大团原子。但通过X射线晶体学（参见第18页）细致观察其结构，我们就能沿着这条链看到它怎样扭曲，把自己挤成一个紧缩的球体。蛋白质化学家有时会用一种别样的方式将这条分子链形状清晰地表现出来（如图11）。从这幅图上，你能够看出，整体的结构是通过某些特定的重复特征——或称“模体”——搭建起来的，比如其中会出现盘旋状的单元，称作α螺旋；还会出现几段相互平行的分子链，即所谓的β折叠。

39

我们还可以从概念上进一步分解蛋白质的结构。分子链是由一个个有特点的小分子团组成的，就像是穿在一条线上的珠子。这些分子团曾经是一个个分立的分子，称作**氨基酸**。天然蛋白质含有20种氨基酸。在分子链上，氨基酸与氨基酸通过一种叫**肽键**的共价键连接在一起。为了形成这个连接，两个分子都丢掉了一部分多余的原子，而余下的部分——链上的枝节——称作**残基**。这条分子链就称作**多肽**。

40

氨基酸残基组成的任何一条链都是多肽。只要对氨基酸混合物加热，我们自己也能够简单地造出多肽。但是光这样造不

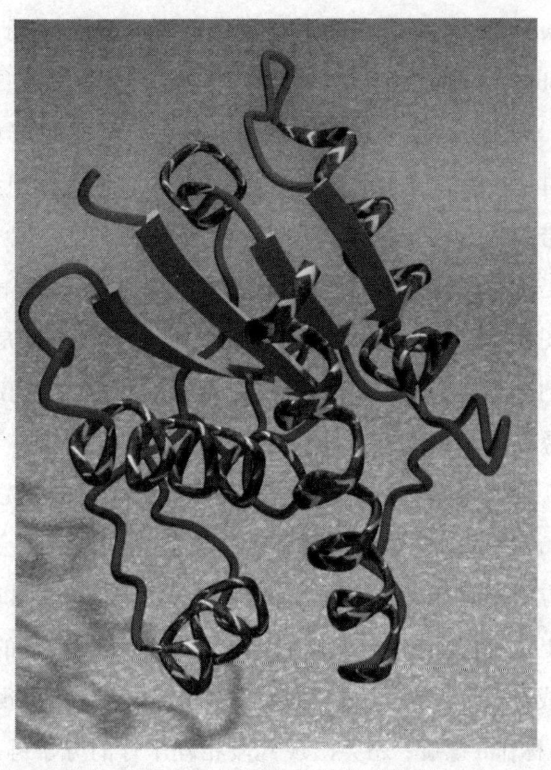

图11　在这种蛋白质分子表示法中，扭曲链的结构被清晰表达出来。图中大体相互平行的几个平直箭头表示β折叠，而标有白色V形的螺旋片段是α螺旋。这个分子是牛体内的蛋白酪氨酸磷酸酶

出蛋白质。蛋白质中，氨基酸在链上的排列顺序——**序列**——并不是随心所欲的。序列是经过选择的（在达尔文的意义上说，就是自然选择），从而保证肽链上的每一部分都处在正确的位置，长链能够在水中收缩卷曲成预定的球状。蛋白质的特殊形状在加热时会被破坏，这个过程称作变性。但很多蛋白质在冷却之后，又会自发地折叠恢复到原先的球状结构。换言之，肽链对它的折叠形状有记忆。

人们对**折叠过程**的细节还未全然知晓——实际上，这也正是分子生物学的一个核心未知难题。不过，我们倒是知道，是什么能够保持蛋白质分子中多肽链的这种紧缩形式。分子链中很多部分可以相互形成微弱的键，称为**氢键**。比如，正是氢键将肽链粘出了α螺旋和β折叠。链上还有一些部分是通过硫原子间的强键结合起来的，这些硫原子来自半胱氨酸的残基。有些残基较难溶于水，于是就倾向于在球状蛋白质的内部结成一团，被链上可溶于水的部分包围起来。因此，由此产生的折叠结构依赖于不同残基的特点，依赖于不同残基在链上的位置——换句话说，即依赖于序列顺序。可以说，蛋白质分子造出来时就带有了它们各自折叠方式的指令。

制造蛋白质时，细胞中的工厂是怎样"知道"该如何排列氨基酸次序的呢？此时DNA（脱氧核糖核酸）就走进了我们的视线。所有细胞中都有DNA分子，体内各种蛋白质的序列就由DNA编码。蛋白质在做所有的工作（或者说大部分工作），而DNA则只是消极被动地等着被解读，一旦需要某种蛋白质的时候就要用到它们了。

DNA信息是用与蛋白质信息不同的另一种语言所书写的，但细胞能够互译这两种语言。DNA是另一种由小的分子单元组成的序列——另一种**聚合物**。但它的组成单元与蛋白质完全不同——并非氨基酸，而是称作**核苷酸**。DNA像一座记录着蛋白质结构的图书馆，由表示**核苷酸序列**的字符书写而成（参见第135页）。粗略地讲，制造蛋白质的信息就编码在DNA的草图里，我们称之为**基因**。

显然这里会牵涉到奇妙的协作。无论是哪里需要酶来完成一项工作，这个信息都必须传递到DNA所处的区域。在人体细胞中及所有生物体的细胞中——除了最“原始”的单细胞的细菌以外，DNA所处的核心隔间称为**细胞核**，细胞核通过自己的膜与外界隔离开来（如图12）。凡有细胞核的生物就称作**真核生物**。

在人体细胞中，DNA打包成一捆一捆的，称为**染色体**。制造蛋白质时，包含着相应基因的DNA就会展开以供阅读。实际上，蛋白质制造并不在细胞核中进行，而是在名为内质网的隔间里，这是个由膜通道所组成的错综复杂的网络。基因首先**转录**成一种与DNA相关的分子，名叫RNA（核糖核酸）。RNA分子从细胞核转移到内质网中，在这里**翻译**为蛋白质。继而蛋白质被输送到需要的地方。因此，细胞里的分子必须具有通信和运输的能力。

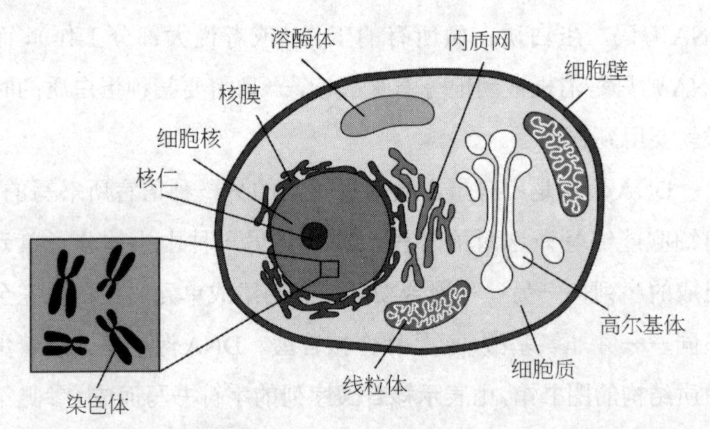

图12　人及其他真核生物的细胞将遗传物质（DNA）分隔在中心的细胞核中。其他不同隔间（细胞器）也各有不同功能，比如合成蛋白质、产生能量等。请注意，仅当细胞准备分裂时DNA才会打包成染色体（如图中所示），其他时候则保持松散的长链状

这整个过程是受到调节的，保证蛋白质不会随随便便就造出来，只在需要时才制造。如果细胞总是任意地制造各种蛋白质，那它很快就会拥堵了。法国生物化学家弗朗索瓦·雅各布和雅克·莫诺在1960年代提出了一项重大发现，解释细胞怎样维持各个隔间之间的秩序。他们表明基因之间能相互调节，可以通过自身编码的蛋白质为媒介来开关其他基因。例如，有些编码了细胞所使用的蛋白质的基因（称为**结构基因**），同时也伴随存在着编码了抑制蛋白质的**调节基因**。调节基因打开时，抑制蛋白就合成出来，绑定到结构基因上，阻止结构基因的“表达”（转录并翻译成蛋白质）。雅各布和莫诺将这些受调控的扩展DNA称为**操纵子**。他们从一个角度展示了细胞中不同基因和蛋白质拥有相互作用的网络。分子生物学家如今几乎解码了人类DNA的全部核苷酸序列，但是对于它所编织的这个巨大网络，人们才读懂了很少一部分。

42 43

基因进化

关于基因的分子科学可谓打开了潘多拉之盒。它不仅帮助我们解释了关于生命的最深层的谜题，还在人类行为以及伦理学上提出了挑战性的难题，并为我们带来了具有争议的新技术。它还颠覆了我们对进化的理解，颠覆了对于我们从何处来这个问题的认识。

基因是遗传特征的流动，它来自我们父母的馈赠，谁都无法避免。个人特征会从父母遗传到子女身上，这一思想非常古老而明显，而19世纪的奥地利神父兼生物学家格雷戈尔·约翰·孟德尔则将它表述得更加具体。基于豌豆杂交实验，他提出存在着调节遗传特征的“颗粒状因子”，这些因子能够从先祖的细胞传递到后代。很快人们就明白了，这种后来被称作基因的“因子”，在本质上就是一种分子。但在20世纪前半叶，很多科学家认为它是蛋白质分子。我们已经看到，霍尔丹也相信这种说法。直到1953年，弗朗西斯·克里克和詹姆斯·沃森推断出了DNA的结构，人们才相信DNA而非蛋白质是遗传的分子物质，是基因的材料。

如此一来，进化论就有了坚实的分子基础——受精卵从父母身上获取的并不是一个预定形的身体，而是一套规划着身体如何发育的基因指令。由于基因的分子构成在一代代变化，进化中的变迁也就慢慢发生。当细胞分裂时，DNA会复制一份，但这并不能总是完美地实现。于是，孩子从父母身上得到的DNA就可能会是带有一点瑕疵的双方基因融合体。这些瑕疵通常不会有什么影响。有时它们会有害（但要注意，大多数遗传疾病是由孩子**继承**了缺陷基因造成的，而不是由于基因复制的随机差错而获得的）。极稀少的情况下，基因突变可能会产生有益的影响，使生物体更适于生存。这种优势可能极其微弱，但微弱的优势也能在繁殖中获得较大的成功比率，从而缓慢提高突变基因在群体中的数量，进化就这样微小地一步一步前进。

这就意味着，基因乃是进化的分子记录。人类和兔子共同的祖先拥有着相同的一套基因。而现今人类与兔子全部基因（即**基因组**）的差异正反映了基因突变逐渐积累形成的分化。这使得科学家能够从基因的分子结构中重建进化史——推断出物种分化的次序。过去古生物学家必须依据生物的身体和骨骼的形状来推断，现在则有了衡量进化中的变迁更轻松、更易量化的分子手段。

具体而言，通过比较线粒体（如图12）这一细胞隔间中的DNA，人们就可以重建进化树，或者称**种系发生学**。线粒体是细胞的锅炉房，能量就在这里产生（参见第四章）。与细胞核中DNA不同的是，线粒体DNA是直接从母体获得的。细胞核DNA会融合来自父母双方的基因，从而发生改变，线粒体DNA的变化则只来自一代代逐渐突变的积累，因而也更好地记录了进化中的变迁。1987年，加州大学伯克利分校的艾伦·威尔逊和他的同事一道，比较了来自许多种族群体的人类线粒体DNA。得出平均突变率后，他们推测出所有样本（也就可以推广为所有人类的线粒体DNA）都派生于同一个版本——20万年前一个非洲女人的细胞，她就是我们所有人的共同祖先。分子中所包含的历史记录如此丰富，任何陶罐碎片或古代墓葬中所含的信息都无法与之相较。

RNA的世界

所有的生命有机体——从最卑微的细菌到最尊贵的国王和王后，都将遗传材料包装在DNA中，并通过蛋白质来发挥作用。这也就暗示着所有的生命有一个共同的起源。[①]最原始的单细胞生物必然包含着蛋白质和DNA，正与现今"简单的"细菌相仿。

但是在那之前是什么呢？DNA编码蛋白质，蛋白质帮助DNA运转并复制，即便在细菌的体内，分子间共生关系也如此

---

① 我们并不知道DNA–蛋白质组合是否是生命的必要条件，不宜草率地断定生命一定没有其他的分子基础。但其他可能的分子基础迄今还没有发现。其实不难想象，对DNA作一些轻微的系统性修改就可能创造出另一种基因系统，但还没有哪种生命体表现出这样的情形。——原注

纷繁奇妙。无论是DNA还是蛋白质，结构都太过复杂，太难随机组装，在早期地球的海洋和湖泊里不太可能从零星分布的有机分子碎片自发形成。要理解38亿年的进化史里最早期的细菌或藻类怎样演变成今天的生物体，这尚属简单（至少在原理上）；而要理解大约短短几十万年时间里地球怎样从不毛之地演变成生命的摇篮，这可就难得多了。

化学家们设想了很多种原创性的方案来解释，他们认为甲烷、二氧化碳、氨、水及氮气等地球早期无机组分可能转化成了制造生命所需的氨基酸和糖。不过这些方案都是探索性的，没有哪一种理论能在生命的化学起源问题上占据优势。然而，岩石、气体和水怎样转化成生物分子原型在概念上也还不算最难，更难的是，这些组成单元是怎样演变成填满了DNA和蛋白质并且运转起来的细胞。这是一个鸡和蛋的问题：若不是两者共存，单有蛋白质或者单有DNA都全无用处。

解决这个谜团的一个好办法是，转而关注两者之间的不起眼的RNA。RNA负责将遗传信息携带到合成蛋白质的地方去。与DNA相比，RNA是个多面手。在1980年代，人们发现RNA可以在自身的重排过程中充当催化剂。人类基因里面有很多“废话”，在清楚地阅读之前需要先切掉它们（参见第140页）。这些“废话”会被复制到RNA上，但在RNA翻译为蛋白质前会被剪掉。这项编辑工作多由酶来完成，但也有RNA分子能够独立完成。这些RNA称为**核酶**，显示它们具有酶的倾向。

在1990年代，生物化学家大大扩展了对RNA能力的认知。他们使用了操纵和改写DNA的生物技术，合成出各种各样的RNA分子，能用来引导各种各样的化学过程，比如将核苷酸连接起来，或者在碳原子间成键。这些研究揭示了，原则上RNA相当万能，足以引发生命起源所必要的化学反应。简言之，RNA既可以是基因携带者，也可以是工人。

因此，不少研究生命起源的科学家提出假说，认为在蛋白质与DNA交互作用诞生之前，存在一段称为**RNA世界**的时期。在与原始地球相仿的条件下，要制造RNA分子同样困难得可怕。不过RNA世界已经打破了蛋白质与DNA相互依存的困境，也概念化地连接起小有机分子的形成和首个原始细胞的出现。 47

合成生命

如果我们最终理解了生命的起源——不一定了解事实上怎样起源，但至少知道**可能**怎样起源——那我们是否能在实验室中重现呢？我们能否从零开始制造生命呢？

对于生命的分子基础，人们已经了解得相当多，研究者完全能够猜测出要怎样建造出人工细胞。这听上去似乎是个有点吓人的前景。如果我们恰好造出一个细胞，它在自我复制方面比“天然”细胞远为出色，那该怎么办？它们会不会像外星人入侵一样殖民我们的星球？

这并不是科幻小说。实际上我认为，无论是好是坏，首例合成细胞会在21世纪内制造出来。如今合成DNA、改写基因已经是非常常规的工作了，很多化学家正致力于从头制造人工设计的蛋白质。生物化学家杰克·绍斯塔克、戴维·巴特尔和皮尔·路易吉·路易西曾经提出：

> 定向进化和膜生物物理学的进步，使合成简单的生命

细胞即便不是可预见到的事实，也已经是个可以想象的目标了。

他们提议，可以通过特制的核酶来构建“极小细胞”。已有学者报告，初步制出了能够组装RNA的核酶（因此RNA也就有潜力自我复制）。人们可以把这样的RNA像普通细胞一样包裹在人工的膜中，但这种膜还需要具备能够增长和分裂的能力，而路易西恰恰造出了这种“可复制膜”。这种能够自我复制的“原细胞”可能会演化出能够把氨基酸组装成蛋白质的RNA分子。研究者称这就使得我们能够“重放早期演化的录像”。

有人预感到这种实验会被用于邪恶的目的。他们应该记[48]住的是，致命的化学和生化武器已经比较容易制得，相比之下用这种手段来开发致命武器要困难得多，选择它很荒唐。不过我们也不可能知道这样的研究会走向何处。这正是分子科学的现实：它是个创造性的学科，最后总能够给予我们努力得到的实物成果。这一路上充满了艺术，充满了奇观，充满了危险，而最终得到的只是我们应得的分子。[49]

第三章

承载压力：由分子而来的材料

空间旅行最困难的部分（除了无聊和危险以外）在于出发。在虚空的空间中，没有太强的引力作用，一次小幅的推进就能使火箭移动几乎无穷远。因此，火箭装载的大部分燃料仅仅是用于让火箭逃离地球的引力。这些燃料和燃烧它们的发动机就占据了火箭总重的大半。我还清楚地记得，“阿波罗号”在离开地球时高大得像座耀眼的大厦，返回来时却小得像鼻尖似的。

因此，要是我们可以从地球的大气层外发射飞船，那负荷就能大大减少了。亚瑟·C.克拉克在小说《天堂的喷泉》里提到过有可能怎样实现。他设想了一种空间电梯：在对地静止的卫星轨道上置一平台，然后用超强的长缆绳将它与地面连在一起。空间设备和乘客首先通过穿梭电梯从地面输送到平台上，然后从平台发射到太空中，这样所需的燃料比起从地面发射就只是一小部分了。

要把绕地运行的平台与地表相连接，我们就需要特别强劲的缆绳，比现今已有的任何一种东西都要强劲得多。这种缆绳

50 还要很轻，要是用普通钢缆就太重了。

其实，要解释我们为何需要具备结实、强韧、耐腐蚀、轻便等特性的材料，并不非得借助于空间电梯这种幻想的概念。但这样的情景能够激发我们思考，到底可以将材料的特性推进到何种程度。我们还可以考虑将超强劲“空间绳索”用于太空发射，先把载荷用缆绳连到绕地卫星上，然后像弹弓一样发射出去。这样的绳索要又轻又结实。即便是在非常普通的需求中，结实的缆绳也求之不得：比如悬吊桥梁，以及在海底固定钻井机。

然而，强韧纤维一直都在我们身边，我们很幸运地接受了大自然大量的馈赠：丝、麻、木材、毛发。在革命前的俄国，人们用蚕丝来防弹。如今为了相同的目的，人们研发出了人造蚕丝。

塑料时代正式开始于20世纪初，这个时代除自然纤维以外，人们又有了合成纤维作为补充，它既有优点也有缺点。最初的塑料是在不断地试错中发明出来的；现代的塑料与此相反，是依据应用场景从分子层面设计出来的。这一章将从分子的视角来看材料，我会把讨论内容同时集中在自然纤维和人造纤维上，因为它们提供了特别美丽且图像化的例子来说明，分子结构如何影响工程师所忧心的材料特性。

不过分子科学和材料工程的互动远比这小小一面要广泛得多。人们可以设计分子，让它们转化为超硬的耐高温陶瓷材料。航天工程、轮机制造和发电中会用到这种材料。人们还可以设计分子材料（尤其是聚合物材料），让它们导电，捕获、传输及转化光脉冲，过滤其他分子，或者保护界面不受污染和侵蚀。其中包括根据环境自动发生变化的“智能”材料，可以用作自动的开关、阀门或者泵。分子材料在医学方面的影响更是举足轻重，而且在未来还会日益强大：它为我们提供了义肢、人造器官和组织、药物控制系统、可生物降解的手术缝线，以及监测身体健康状况的传感器。未来有可能分子工程会培育出新的肾脏或心脏，用来移植以取代损坏的脏器。

身体的缆绳

要说我们的身体是蛋白质组成的许多细胞共同体，这一观念其实并不符合我们的体验。我们会感知到身体是各种结构的交织，包括皮肤、骨骼、肌肉、毛发和指甲。这个材质框架具备让我们与周围世界互动的必要性质。皮肤外表只是一个保护层，细胞一旦形成了这样的组织就会按既定程序死亡。起到保温隔热作用的毛发也是这样。指甲同样如此，它们其实是过去能够抓、刺、撕、扯的爪子在进化过程中的残余。骨骼和牙齿主要由坚硬的无机材料磷酸钙组成。这些材料中的大部分都在我们的一生中不断更新，也有一些不更新，如眼球晶状体中的蛋白。

身体中的这些自然材料发挥着机械性的作用，维持身体结构完整。它们就像楼房中的砖块、大梁和外墙，保护工人们免遭坏天气的侵扰，并在其中安置正常经营所必备的复杂电线和水管。身体很多结构组织都是蛋白质。组织蛋白和酶不同，不需要执行那些精巧的化学变化，它们无非得有（例如）坚韧、有弹性或者防水的特性罢了。原则上讲，许多非蛋白质的材料也都满足这些条件，如植物就是用纤维素（基于糖的聚合物）来构建自身组织的。但蛋白质的奇妙之处在于可以千变万化。它的

52

分子链可以编织成强韧的纤维，而交联[①]或交缠起来则可以形成角或者爪子的坚硬基质，或者形成有弹性的薄层。而且，制造蛋白质所需的原材料在细胞中非常充裕。蛋白质是在基因中编码的，所以赋予结构蛋白力学性质的相应分子特征可以精细地调节，整个分子可以可靠地复制。

人体中含量最大的一种结构蛋白占全部蛋白质总质量的四分之一，这就是胶原蛋白。这是一种比较简单的蛋白质，其链状分子中大部分是两种氨基酸：甘氨酸和脯氨酸。分子链中每三个氨基酸就有一个是甘氨酸，中间夹着脯氨酸和别的氨基酸（尤以赖氨酸为多）。中间的脯氨酸单元经过化学修饰，加上了一个氧原子。这一过程有维生素C的参与，这也就是人体需要维生素C来保持组织健康的原因。缺乏维生素C将导致坏血病，病因就是损坏的胶原蛋白没有被替换。

天然的蛋白质结构材料与大多数合成的聚合物塑料有所区别，胶原蛋白就是一个例证。这两类材料都是由链状分子组成的，但在结构蛋白中，分子链经过复杂的排列整合在一起，形成了更厚重的小纤维，就像细线编织成麻绳那样。小单元组织成大的结构，大结构又组织成更大尺度的结构，这种逐层组织结构要素的排列方式称为**层级结构**。建筑师们也学会了利用相同的原理：例如，在古斯塔夫·埃菲尔标志性的铁塔中，有的支柱是由小钢梁的网状结构组成的，而这些小钢梁又有一部分由更小的钢梁组成（如图13）。

53 塔的设计有很多种，胶原蛋白也表现出若干种不同的大尺

① 指线状分子横向连接成网状。——译注

图13 结构层级，以埃菲尔铁塔为例，这种结构在天然材料中十分常见

度结构，但所有的结构都是从基本的小尺度元素构建起来的（如图14）。每条胶原蛋白分子链会卷曲成螺旋状。三条链相互缠绕，形成绳索一般的三螺旋“微纤维”。这些微纤维又通过不同方式相结合。比如，它们可以交错排列形成较粗的股，称为纤维束。由于交错排列，在电子显微镜下就能够观察到它的暗条纹。纤维束组成了细胞之间的结缔组织，就是将我们的肉身束缚在一起的绳索。骨骼中包含点缀有羟基磷灰石（主要成分是磷酸钙）微小晶体的胶原纤维束。因为含有大量的蛋白质，所以骨骼柔韧有弹性，同时又很坚固。

不过，胶原纤维自身并不算特别强劲，因为其中的分子并没有相互连接或者相互交缠。而其他种类的胶原蛋白含有微纤维的交联结构，组成某种坚韧的网络。外层皮肤与内层皮肤相隔离的膜就是由这种结构形成的。

54

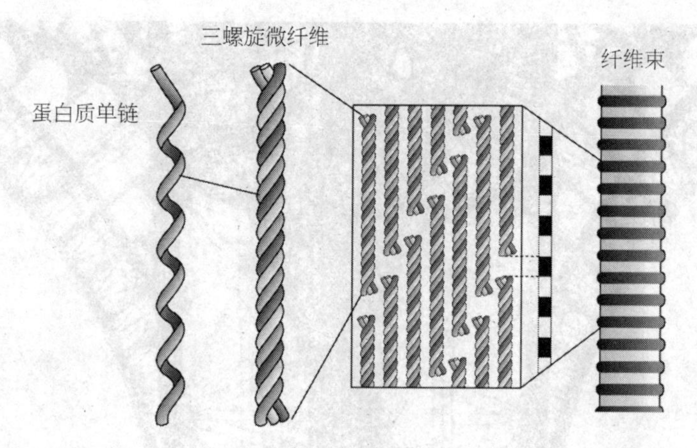

图14　胶原蛋白拥有螺旋套螺旋的层级结构。在显微镜下，胶原微纤维的交错排列形成了暗条纹，其中有金属染色剂附着在微纤维的端点

结缔组织中是无序缠绕的结构，与此相反，眼角膜则由胶原纤维有序地排列堆积。这些纤维太小，不足以散射光线，因此这种材料几乎是透明的。由此我们能看到一条自然界中随处可见的基本设计原理：通过调整化学成分，以及——最重要的——改变相同基底分子的层级排列方式，我们有可能得到多种不同的材料性质。

胶原蛋白还可以组成肌腱和韧带这种强韧而有弹性的结构，并构成坚固的牙质。而我们的体表——如皮肤、毛发、指甲，以及动物的角和蹄，其中的蛋白质则属于另一类。这些组织的主要成分是角蛋白，这是另一种有着层级结构的纤维。角蛋白的分子链同样也弯曲成螺旋状，两条一组卷曲成束：两条这样的分子束再次相互缠绕形成“超螺旋”，称为原纤维；八条原纤维结成一簇，就形成了角蛋白的基本绳索。类角蛋白的蛋白质通过硫原子交联，形成不规则的阵列，包裹住角蛋白纤维，就像混凝土把钢筋包裹起来那样。而交联决定了材料的硬度，比如毛发和指甲中的交联就比皮肤更为紧密。通过破坏硫原子的交联作用，我们可以让头发变得更柔顺，于是就可以将卷发拉直。

毛发是一种很有用的天然纤维，但大多数胶原蛋白和角蛋白组成的材料并不是线状的，而是薄层状（比如皮肤）或者块状的（比如角或蹄）。在分子尺度和显微镜尺度上，这些材料都呈纤维状结构，其原因在于，对于单细胞制造体系而言，造出这样的结构比造出其他结构——比如铸造固体块——要来得简单。但在这之后，生物体就将微纤维组织成了其他的形状。

网之梦

另一方面，丝这种材料则表明当生物真正需要时，进化会以多么深刻的方式去应对制造纤维的挑战。这种物质要能够形成线状，对飞过的飞虫要几乎透明，而且要有足够的弹性、能够吸收飞虫撞上蛛网时的能量，还要足够强韧、在飞虫撞击下不会破损（如图15）。无论是与钢铁还是与最棒的人造纤维相比，丝线都更加强韧。工程师们为丝绸的韧度所折服，纺织业者则赞叹它那奇妙的光泽、漂亮的纹理和吸收染料的能力。

蜘蛛出于多种目的而吐丝，不同目的下吐出来的丝也有所不同。蜘蛛织网时先从牵引丝织起，再用其他丝做支撑纤维，将蛛网连接到树枝或房梁上，还要造捕捉猎物的丝线，以及用来包裹发育中幼虫的丝线，等等。所有这些蛛丝都是由蛋白质链组成的，其中的氨基酸主要是甘氨酸、丙氨酸和丝氨酸，但具体配比则需要根据用途作些调整。

56

蛛丝之所以有神奇的特性，正是由于其蛋白质链特别的组

图15　蛛丝是目前已知的强度最高的纤维状材料之一

织方式。在胶原蛋白和角蛋白中，长链都弯曲成螺旋状，而在蛛丝中，基本的组织结构要素则并非螺旋，而是片层。相邻的链会整齐地并排放好，彼此通过较弱的氢键（参见第41页）连接起来，这样就将长链压缩成了β折叠（如图16）。

规整且较硬的片层结构可以一层层堆积起来，形成微小的三维蛋白质微晶。在蛛丝纤维中，这些微晶非常小，各方向长仅约五万分之一毫米。在微晶区域以外，蛋白质链会继续延伸、形成不怎么整齐的区域，彼此纠缠。所以，蛛丝其实是复合的材57料，微小的晶体散布于更具弹性的蛋白质阵列中——和骨骼有点像，只不过晶体不是无机物，而是蛋白质本身。

丝蛋白的微晶区域通常包含规律性重复的氨基酸序列。比如，在家蚕的茧中，甘氨酸—丙氨酸—甘氨酸—丙氨酸—甘氨

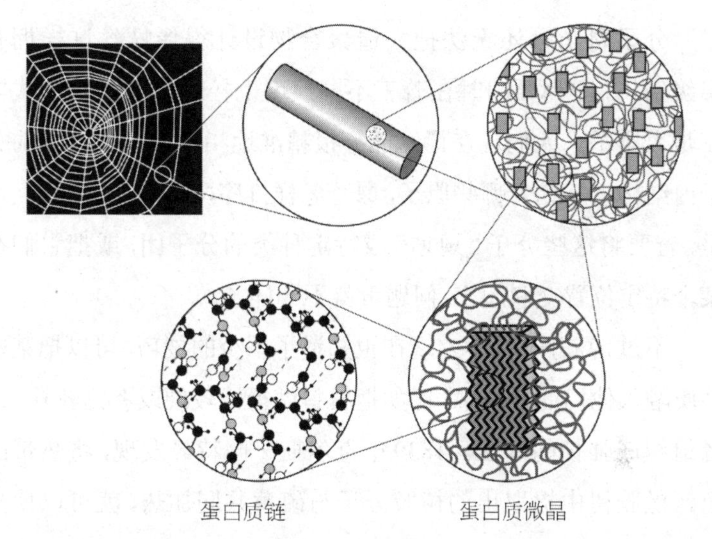

图16　蛛丝的层级结构中包含好几层。分子尺度下，平行的分子链组织成整齐的（类晶体的）β折叠。虚线表示氢键

酸—丝氨酸的序列就在链上重复。而在无序区域中，氨基酸序列也不规律。

丝纤维在水中不能溶解——要是蛛丝会溶解在朝露之中，那也就没什么用了。不过蜘蛛却是从溶有蛋白质的水溶液中吐出丝的，看似神奇地将可溶的分子转变成了不可溶的。这是因为分子链组织的方式发生了变化，所以溶解性改变了。蜘蛛在丝腺中制造丝蛋白，丝蛋白在这里仍是可溶的。溶液会从丝腺传递到吐丝器中，在这个过程中，溶液浓缩脱水，分子链也开始在氢键的作用下折叠起来。当蛛丝离开吐丝器时，其中大部分的水都已经被挤出了，分子链形成β折叠。一旦在微晶区域分子紧密地堆积起来以后，水分子就很难再渗透进入，于是丝纤维基本上就成为了不可溶的固体形态。

分子科学家还无法把人造聚合物设计得像蛛丝这样拥有层级结构，因为要同时在若干不同尺度上控制分子堆积方式实在是太难了。人们现在已经能够很精准地指定合成聚合物的分子内结构——包含哪些原子，原子怎样排序，空间如何排列。可是，若要将这些分子“规划”成特定种类的分子团，或把它们在某个特定位置交联起来，问题可就不一样了。

不过，高分子化学家现在也知道了不少的技巧，可以把某些性质融入他们的材料中，比如把材料做得比较硬或者比较软，或者组织成强韧的纤维。1839年查尔斯·固特异发现，将热带的巴西橡胶树中提取出的树胶分子与硫黄共同加热，就可以使分子交联。这种称为“硫化”的过程将柔软的树胶转变成有弹性的形式，也就是我们所说的橡胶。

天然树胶主要由一种烃类聚合物构成，称作聚异戊二烯，分子链中只包含相互连接的碳原子和一些附带的氢原子。当今很多大规模生产的塑料也都是烃类聚合物，如聚乙烯、聚丙烯、聚苯乙烯等，它们都是炼油的产物。但卡车轮胎所用的橡胶却是59天然产物，因为要去合成它仍然太困难。

在20世纪中期以前，要将烃类小分子连接起来形成聚合物长链可是件靠运气的事情，结果会产生很多种不同类型的分子链：有的短，有的长，有的会分叉，有的是直链。因为链的结构会决定它们怎样堆积，而堆积的方式又进而决定了宏观材料的性质，所以对合成缺乏控制就意味着，高分子化学家们很难精细地调节材料性质。大自然能够通过精致的结构控制用同样的原材料制成好几种明显不同的纤维，而化学家却只能每次都得到相同的塑料。在1950年代，人们研制出了特别的催化剂，对分子链结构能够进行更深入的控制，从而控制分子堆积的方式。举个例子来说明它的意义：对于聚乙烯，我们不但能造出过去那种柔软、低密度的形式，还能够造出新的坚硬且高密度的形式，扩大了使用范围——比如用来制造包装用的不易弯曲的桶和瓶，以及工程和建筑领域所用的管道和板材。

这两种形式的区别在于，它们分子链堆积的整齐程度不同——也就是说，它们的结晶程度不同。蛛丝告诉我们，结晶程度越高，材料就越坚硬、越强劲，密度越高，越难溶于水。对于强韧的纤维来说，这些都是我们想要的性质。分子链堆积得越紧密、越整齐，胶着力就越强，材料的硬度也就越高。

于是，制造高强度聚合纤维的一大主要挑战就是提高分子的排列有序度。对于聚合物，只有由直链（即没有分叉）分子构成才有可能按分子链整齐排列。但在蛛丝中，分子的堆积则是由于它们倾向于“压缩”成结晶态β折叠。换言之，蛛丝分子链自身就带有某种“排列指令”的程序。我们能够像这样制造人造聚合物吗？

我们的确可以做到。蛛丝蛋白质排列整齐，是因为不同分子链的单元之间存在吸引力，使链像拉链一般锁定。而芳纶这类合成聚合物（杜邦公司著名的凯夫拉纤维的制造原料就是它）的链之间也存在着相似的作用力。

若要制造约束空间平台的绳索，凯夫拉纤维是目前最好的备选材料之一。它的抗拉伸能力比钢要强，但它比钢轻得多。人们用这种纤维做橡胶轮胎的强化帘子线，做防弹衣的织料，做航空航天工程领域高强度复合物中的加强材料，甚至做石油钻井平台的锚定缆绳。

但实际上蛛丝比凯夫拉还要强韧。蛛丝内部分子链的高度有序并非仅仅因为分子在溶液中倾向于压缩起来。当蛋白质溶液从丝腺经通路流向吐丝器时，聚合物分子链会沿着液体流动方向对齐，就如同风一吹，我们的头发也会排齐。随后，在蛛丝从吐丝器喷出的过程中，也会产生同样的作用。这种效应称为剪切诱导对齐（因为流动液体受到所谓“剪切应力”的作用）。

所以，即使我们有可能人工制造类丝蛋白质，要用它们造出丝线也完全是另一码事。如果我们机械地让丝蛋白溶液（天然或人工均可）喷射而产生纤维——就像胶水从瓶子喷头中挤出来那样，结果得到的纤维还是不像真正的蛛丝那样坚韧。有些研究者相信，除非造个微型机器来模拟蜘蛛的吐丝器，否则无法造出可与天然材料相媲美的人造丝。

剪切诱导对齐已经被用在工业流程中，用以将聚乙烯制成极强的纤维。这种纤维是通过复杂的流程从一种胶状物质中提取而来的，经由此流程，聚合物分子链排列高度整齐有序。人们有时称它为“火箭丝”，它比凯夫拉还要结实，而且和很多有机

材料不一样的是，它的化学性质非常稳定。这样的性质使得它很适合用作长效的外科缝合线。

材料的基因

制造人工的丝状聚合物的一种办法是，使用生物技术将产丝的基因植入细菌体内。和其他蛋白质一样，蛛丝也是由DNA进行基因编码的：分子链上的氨基酸序列正是由蜘蛛染色体中的相应核苷酸序列决定的。换言之，蜘蛛拥有了制造这种聚合

物的分子设计图。（不过请注意，因为吐丝过程很复杂，单凭设计图纸并不足以造出优质的蛛丝！）

生命有机体能够完全精确地指定一个聚合物的分子构成，这种能力让人羡煞。现代合成技术已经能让化学家在很大程度上掌控分子链的构成，比如他们可以通过把一种聚合物支链嫁接到另一种聚合物主链上，或者通过对调单个分子链上的两块不同部分，来制取杂化聚合物。但是，要想合成一种包含多达上千个单元的聚合分子，各种单元反复出现，顺序相当随意、复杂，此外还要保证材料中的每条分子链都一模一样，这种要求就远远超出了我们现有的合成水平。用语言学来打比方的话，当今最先进的合成分子看起来有点像这个：aaaaaaabbbbbbbaaaaaaabbbbbbbaaa……而大自然中的聚合分子则更像是我现在正在写的这整句话，其中蕴含着意义。

借助生物技术，我们可以从一种生物体的大段DNA中剪出一小段，再把它粘贴到另一种生物体的DNA中。然后，受体就会把新的基因当作它自己的（如果顺利的话），通过转录和翻译机制把相应的蛋白质制造出来。这种生物技术中具有潜在价值（我想也是争议较少）的一大方面，就是将人类的基因转移到细菌体内，之后在发酵池中培养细菌，制造医用蛋白质。不过有些合成科学家还意识到，这也是一种制造蛋白质材料的办法。

62

我们可以将真正的蛛丝基因转移到细菌体内，[①]我们也可以用这种方法“书写”并表达人工基因。结构蛋白中常包含重复的氨基酸序列，因为它们形成的纤维需要沿伸展方向保持各

① 有一项正在进行的研究工作，是将产丝基因转移到山羊体内，期望从羊奶中获取丝蛋白。——原注

处均匀。要造出能合成循环序列的基因相对容易一些：只要先造出一个重复单元，然后再把若干单元合并起来就可以了。（要保证得到的合成基因中包含固定数量的重复单元也有办法。）一些研究人员已经开始利用这种思想来制造基因工程蛋白质材料，对材料分子链的特定结构进行定制。“人工设计”的类丝材料就是用这种办法制造的，别的类似于胶原蛋白和弹性蛋白（皮肤中一种有弹性的蛋白质）的蛋白质相关合成材料也可以这样制造。我们还可以设想一些杂化材料，用天然蛋白质（比如某种酶）制成人工设计的类蛋白质长链材料：比如，将蛋白质堆积成不溶于水的β折叠微晶体，形成防水的外层。在医学应用方面，这种类蛋白质材料可能的优势在于，它具有生物相容性及生物可降解性。

细胞的骨架

细胞中有纤维组成网络结构，纤维强度远高于胶原蛋白和角蛋白这样的绳索状结构蛋白。这些纤维称作微管。它们是杆63 状的轨道，分子引擎就沿着微管围绕细胞运送包裹。微管还是一座脚手架，细胞可以基于它来变换形状——比如让变形虫伸出伪足。在我们的呼吸道中，像汗毛一样的纤毛从细胞中突起，能推动过滤灰尘的黏液；对细菌来说，像鞭子一样的鞭毛能够在液体中驱动细胞螺旋状的运动。这两个例子其实都是微管。

顾名思义，微管是空心的，呈管状结构。它们是由一种叫作微管蛋白的蛋白质组成的。这种蛋白质并不是纤维状的，而呈紧致的球形。它包含两种几乎完全相同的分子，两个分子结合成哑64 铃状，哑铃又像砖块一样堆起一座圆柱形的烟囱（如图17）。

图17　微管是由微管蛋白组成的细丝,它构成了细胞的脚手架

一个个微管蛋白单元可以附着在微管的端头，也可以从微管端头上分离出去，也就是说这种纤维可以加长或者缩短。变形虫可以缩回它的“腿”，只要把拉动这条腿的微管解散就可以了。因为微管容易组装，所以它就在细胞分裂中扮演了核心的角色。一旦要分裂的细胞复制好染色体，它就开始组织两簇放射状的微管，称为星状体。来自两个星状体的微管相遇时，它们就在末端融合，两个星状体焦点之间就形成一束桥接的纤维，称为纺锤体（说到纺锤自然会想到丝，而细胞分裂就恰恰称为有丝分裂）。染色体附着在纺锤体上，两边分别向两极拉伸，将染色

图18　纺锤体由微管组成。在细胞分裂期间，染色体（图中部可见）基于纺锤体的架构进行排列和整理

体拉成两半（如图18）。通过这种方式，纺锤体结构保证了复制的遗传物质分成完整的两组。接着按照微管的架构，细胞被拉65分为两半，每一半都有一套完整的染色体。

抗癌药物紫杉醇（参见第29页）之所以发挥作用，就是因为它破坏纺锤体的组装，从而抑制癌细胞的增殖。它阻碍了微管的分解，而在两个星状体盲目地相互搜寻时，微管分解正起着关键的作用。可惜紫杉醇对健康细胞也会起到相同的作用。不过癌细胞分裂速度要迅速得多，所以它们受到的影响最大。

碳纳米管

在1990年代以前，几乎没有人认真地考虑过用管状分子制造高强度的合成纤维。将小分子结合成长链十分简单，但要排布成管子看上去就困难多了。

但到了1991年，在筑波NEC公司工作的日本显微镜学者饭岛澄男发现了一种管状分子，足以成为人们所知的最强韧的纤维。这种结构称为碳纳米管，是碳的原子蒸气自我组装形成的。

饭岛当时研究的技术其实是富勒烯这种碳分子的制备，这种分子包含几十个碳原子，结合在一起形成空心的笼子。第一个富勒烯是含60个原子的笼子，称作巴克敏斯特富勒烯，它是在1985年发现的。它奇怪的名字得自美国建筑师理查德·巴克敏斯特·富勒，他利用六边形和五边形的面建成建筑物的圆顶结构，引领了穹顶建筑的潮流。巴克敏斯特富勒烯——或称C_{60}——恰在分子尺度上与建筑物的穹顶有着相同的结构。它用六个或五个碳原子先分别组成六边形或五边形，然后再连接成球形的笼子（如图19）。

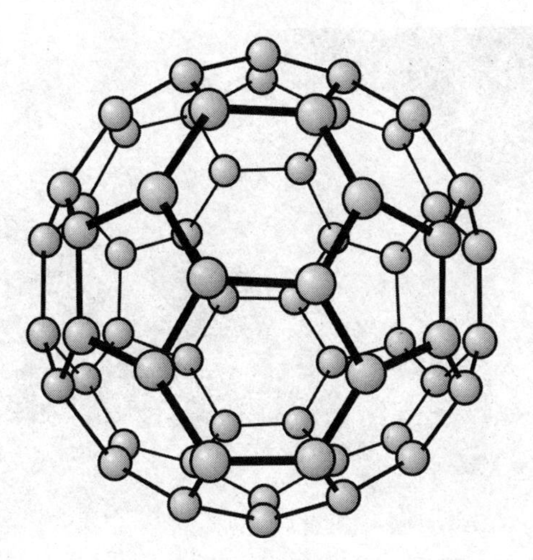

图19　C_{60}分子呈近似的球形，由五边形和六边形碳原子环组成

1990年，人们首次发现了大量制造富勒烯的方法。这种办法要在两根石墨棒（其中全部是碳原子）间放电。电火花的能量能够使一些石墨转化成蒸气，蒸气冷却时碳原子就重新结合，形成C_{60}和其他的碳笼子。而饭岛当时使用了略微不同的条件去生成富勒烯，在电子显微镜下检查乌黑的残渣时，他发现了一些新东西。

他发现碳渣里面全都是针状的物体，直径只有几个纳米。再仔细地调查，他看到它们其实是碳原子组成的空心圆柱，每根针状物都包含若干圆柱，一层套一层，像俄罗斯套娃那样（如图20）。这些物体后来被称为碳纳米管。之前从没有任何人想到——即使在发现富勒烯后也没人预料到，碳原子竟会自发地以这种方式进行组织。

碳纳米管的壁面是碳原子以六边形构成的片层。这和石墨的结构一模一样，只不过纳米管的片层是弯曲的，卷成圆柱。在

图20　碳纳米管的尖端，电子显微镜下的横截面图像

管的端头，片层会扭曲形成平顶的小帽。我们一般会认为石墨是种脆弱的材料——毕竟在纸上拖动它就能擦下一部分黑色的 67 碳，所以我们才用它来做铅笔。可它之所以脆弱，是因为碳原子片层间的连接很松弛，各层之间能够滑动。而对于其中的单层，原子间的连接则很牢固，理论预测这一层分子的强度很大，堪与金刚石相媲美。当这些石墨状的原子片层通过化学键局部结合，材料的强度和刚度都会得到很大的提升，就像传统的碳纤维那样。

碳纳米管大概是终极的碳纤维了，这种管子正是石墨状的碳原子片层的无缝连接。理论预测，碳纳米管的抗拉性能比金刚石还要强，也比凯夫拉、蛛丝或者其他任何你能想到的自然的、人工的纤维都要强。富勒烯的发现者之一、在休斯敦的莱斯大学工作的理查德·斯莫利认识到了这一点，他提出，如果人们真的要建造空间电梯，那么碳纳米管可以把它固定在地球上。

不过这其中还有障碍。迄今为止，碳纳米管生长的长度还远远不到一毫米。这种绳索实在不怎么长。这也使得测定它的真实强度很难，尽管人们在微观尺度下做了一些实验，验证 68 了这些管子确实非常强韧、坚硬。（能否与金刚石相比？人们还不确定。）

要利用碳纳米管制造可用的绳索，我们就需要研发一种控制它们生长的办法，确保它们能无限地延伸。在普通的聚合物合成中，有一种技术称作“活性聚合”，它可以使我们随心所欲地抑制并重启分子链的生长，于是越来越多的单元可以附着在分子链上，使链不断变长。如果有人能对碳纳米管研发出类似的过程，那将会是革命性的进展。但人们目前对碳纳米管如何生长仅有一点模糊的认识，更难以搞清如何控制它。[①]就目前而69 言，空间电梯还需要我们慢慢等待。

① 位于佩萨克的波尔多第一大学的法国研究者菲利普·普兰和他的同事们报告了尝试性的第一步。他的课题组用这样的办法来制作碳纳米管的纤维和带状物：在类肥皂分子的作用下，让碳纳米管悬浮在水上，然后注射到黏性的液体聚合物中。从毛细管注射时，碳纳米管会排成一条线，就像蛛丝蛋白在从吐丝器射出时能够排列整齐那样。排好的纳米管彼此粘合成纤维，继而可以干燥并进一步处理。不过这种链状物并不是由**连续不断**的碳纳米管组成的，因此并不像传统的碳纤维那么强劲，也就更比不上金刚石了。——原注

第四章

燃烧：分子与能量

想象一下，如果汽车像人体一样，最快速度只能在短短的冲刺阶段维持一小会儿，那会怎么样？那我们就不可能沿着高速公路开到时速（不妨让我们说实话）80英里了[①]——最快的速度只够我们开半英里去商店而已。要走的路程越远，速度就会越慢，这样才不至于把可怜的车子跑坏。（没准你跟我一样，也开过这种车？）

表面而言，其实汽车和人类比我们想象的要更具相似性，这么讲的原因可不是弗兰·奥布赖恩提到的那种。[②]两者都通过在氧气中燃烧富能燃料来获取能量，而且都排出二氧化碳。不过汽车能以接近极速行驶很长时间而不疲劳。只要你往油箱里不断加油，它就能几乎无限地一直行驶。而短跑运动员却不能用短跑的速度跑完整个马拉松，即使不断地给他们嚼葡萄糖片

---

① 实际上英国高速公路限速70英里每小时，作者指超速现象很普遍。——译注

② 你要是不懂我的意思，可以去读读《第三个警察》！——原注（在小说中，人和自行车融为了一体。——译注）

也无济于事。这是为什么呢？

这个问题把我们引向了人体供能的分子机制，也就是代谢过程。在很多方面，对生命最好的惯用定义在于代谢，而非复制。进化生物学家也许会说，我们生存的目的在于繁衍——可是，我们却并非时时刻刻都在努力进行生殖活动，哪怕最好色的人也不是。而一旦停止代谢，即使只停下一两分钟，我们可就死翘翘了。

摸摸你的手，你会觉得它是暖的。（要不是的话，就摸摸腋下或者舌头。）我们的身体一般都比周围环境要暖。无论行走还是睡眠，我们的体温总保持在接近37℃的健康温度。要维持它只有一种办法：靠细胞持续输出热量，这是代谢的一种副产物。热量并不是这里的关键问题，它只不过是无法避免而已，因为任何能量转化都必然会像这样浪费掉一部分。我们代谢过程的根本在于制造分子。细胞若不持续地更新自身就无法生存，它们需要为蛋白质制造新的氨基酸，为细胞膜制造新的脂类物质，为细胞分裂制造新的核酸。只要我们活着，细胞的车轮就不能停下，而转动车轮正是要消耗能量的。

因此，细胞的社群颇像我们对工业革命社会的那种老印象：在大规模的社会化生产中，一大批工人专门生产能量。在我们的肝脏中，制造能量是头等大事，里面每一个细胞中都有成百上千个能量工厂。就像威廉·布莱克《耶路撒冷》中所写的黑暗邪恶工厂，它们既制造有用的东西，也制造垃圾。不过细胞整体会有更有效的办法防止垃圾污染自己的卧榻。

既然制造能量对生命至关重要，那么代谢中的任何瑕疵都会令人不安。我们能看到生命是多么脆弱——但凡哪个过程中断了一下，整个系统就会陷于停滞，就像我们的社会结构要依赖于连续不断的电力和燃气供应一样（更不用说清洁的空气和充足的水）。而人体这台机器（请原谅这种启蒙时代的比喻）非常可靠，如果情况顺利能运转80多年之久，直到各部分无可逆转地失效，这正是进化之创造力的深刻证据。 71

火的研究

还是让我们从简单的，至少貌似简单的说起。1850年，英国科学家迈克尔·法拉第在伦敦的皇家研究所开了一系列讲座，题目是“一支蜡烛的化学史”。他希望展现，在蜡烛火苗耀眼的光亮中，我们能够读出整个化学科学（那时人们所理解的化学）和超越化学的更多东西。他这样讲：“要说自然科学研究的入门，没有比这更佳、更人人皆宜的敲门砖了。”

在训练有素的学者眼中，任何一种自然现象都可以是布莱克的那一粒沙，[①]都可以是窥见无垠世界的一扇窗。蜡烛自然也是，尤其是在19世纪的伦敦，那时的展示讲座还无须让人目眩神迷。我虽不太清楚内燃机是怎样工作的，却知道它和蜡烛燃烧的区别并不大。蜡烛燃烧是氧化反应，即某种燃料与空气中的氧气结合，产生热以及——此例中还有的——光。

即使是石蜡的燃烧，人们至今也并未完全搞清其中的所有细节。而当时的法拉第，自然也肯定没有明白问题中的许多重要方面。不过，燃烧的本质其实就是产生能量的一切化学过程的本质。首要条件是，它必须是个下降的过程。

① 威廉·布莱克写有诗句“一沙一世界”。——译注

这是化学变化中的一个关键点，也是宇宙中一切变化过程的关键要素：变化都包含上升和下降这两种方向，而在充分自然的情况下，变化都会朝向下降的方向。而上升和下降的地形是谁决定的呢？归根结底，决定性因素是一种现在略显神秘的东西，称为熵。无人能够反驳的热力学第二定律这样讲：在一切变化过程中，宇宙的熵总和都保持增加。[①]

72

在流行文化中，人们把熵等同于无序。这是个还不算坏的简化：系统的熵所度量的是，有多少种办法能够重排系统中的原子，而又不引起任何可察觉的变化。如果有人进入我的办公室，将本已乱七八糟的纸张重新打散成若干堆，这样的变化我很有可能不会察觉到。但如果像做梦一样，他把所有论文都认真地整理好，排序归档，那我马上就会注意到这个变化。粗略地说，相较于无序系统，有序系统中不可分辨的重排方式较少，因此熵较低。

第二定律其实表达的是这样一件事：系统从有序向无序发展的可能性较大，不过是因为无序的可能数量要比有序的可能数量多。若系统包含好几万亿个分子，这种概率化的陈述就几乎成了确定性的。第二定律之所以称为定律，只是因为违反它的可能性极小，几乎不可能发生。

不过**确实**在某些情况下，事物会变得更加有序。水蒸气冷凝可以成为对称的六角形雪花。宇宙定律也不能阻止我们把一堆砖块排列起来盖成房子。这些情况都是真真切切的，但它们

① 严格地讲，这条定律应该针对不可逆的变化。在能够逆向进行的变化中，熵的变化为零。所谓逆向进行，我并不是指单纯把一些移动的东西放回原位，而是指要复原一切体温的散发、一切空气的移动、一切摩擦力的作用。这也就意味着，把一件物体拿起来这个变化就不可逆。——原注

并未违反熵增加原理。原因是，第二定律对于宇宙整体才适用。我们可以使某一部分变得更加有序，但却要以更大的代价让别处变得更加混乱，以补偿那些有序性的增加。一般这种补偿的形式是热量。我们的确可以通过辛劳工作砌一堵墙，但墙砌好了，我们的身体也向环境中辐射了不少热量，增加了周围环境的无规则热运动，而熵的总账保持盈余。只有增加无序性才能得到秩序。

73

而生命细胞显然无视熵增的压力，维持着自身的组织结构。为了解释其中的缘由，物理学家埃尔温·薛定谔在《生命是什么》一书中提出“负熵”这个有点模糊的概念，讲生命有机体能够从环境中摄取它。这听起来有点像是某种生命活力论乔装打扮成了热力学的样子，而且今天我们还会在言谈中听到有的人将酶称作“逆熵器”。不过，生命的机理没什么特殊或神秘的。再摸一下你的手（腋下、舌头）。你的身体所做的不就是将熵从体内抽到环境中吗？

我前面说到，化学变化都朝向下降的方向。但有时候，变化似乎会表现出朝着上升的方向。酶就特别擅长将分子向上升方向驱赶，而非生物的世界也会发生这种情况。但在这些例子中，方向之所以会逆转，是因为它有更强大的下降反应过程在驱动。你可以将一个物体推上斜面，只要它和另外一个更重的物体通过滑轮连接，而另一个物体在下降就可以了。重物体下降，轻物体上升。化学家彼得·阿特金斯这样讲：“理解……生物化学，本质上就是要在巧妙遮挡的幕后寻找重物。”

生物化学中的能量制造，基本上就是要在单个重物完全落下之前，拉起尽可能多的重量。像我们这样的动物，重物就是食物分子中蕴藏的能量。

我们也可以用河流中的水力发电来打比方。无论我们做什么，河水都是要下降的，它永远不会自发地流回到高山上。于是我们的目标就是捕获下落过程中尽可能多的能量释放。这也正是我们体内的能量工厂与蜡烛火苗间的主要区别之一。火苗中的燃烧是不可控的，我们只能得到光和热。而在人体中，燃烧受到严格的控制，按照一层一层的顺序分步进行，化学能量就在各个步骤间被释放或者储存。

74

燃烧的糖

发电厂会燃烧煤、石油、天然气，但显然它不仅仅只是一个扩大了的燃烧的锅炉。燃烧只是达到最终目的的一个手段。热量用来将水变为水蒸气，水蒸气的压力推动了涡轮机，涡轮机带动电线线圈在大磁臂中旋转，于是电线中就产生了电流。在这个过程中，化学能转化成热能，再转化成机械能，最后转化成电能。所有发电厂都会有大量安全和管理机制：有人工去检查压力仪表，检查运转机件结构是否完好；自动传感器会进行测量；故障应急装置能够防止系统发生灾难性的故障。

细胞中的能量制造也一样复杂。简短的一番描述根本对不住这系统的非凡美妙。细胞似乎把一切事情都考虑到了，用蛋白质器件来精细地调整一切。

细胞的主要投入产物是燃料和氧气，即食物和空气。细胞若是缺少氧气，它们的火苗就会熄灭。我们可能会在食用之前对燃料施以一番精致的烧炙，因为这些初步的燃烧能创造出令味觉愉悦的化合物来。但无论是巧克力棒、菠菜叶还是猪蹄，任何食物都会在之后被分解为更加同质化的燃料。这是在消化时发生的，胃和肠道中的酶能够将烹饪的美味分解为原始的分子组分。

食物中有多种富含能量的成分，本质上可以分为糖类、脂类和蛋白质。糖类是由葡萄糖分子连成长链形成的聚合物。消化过程中，脂类可以分解为脂肪酸分子和甘油分子。同等质量下，脂类包含的能量是糖类的两倍，心脏有65%的能量都是从脂类中获取的。

葡萄糖是代谢过程中的主要“重物”之一。它在酶的辅助下氧化，推动一种名为三磷酸腺苷（ATP）的高能分子形成，这种分子就在细胞的其他过程中充当能量来源。ATP是生物化学的能量包。很多酶催化的反应都需要ATP来推动反应往上升方向进行。[①] ATP是维护细胞完整及组织结构的关键，所以细胞要通过燃烧每一个葡萄糖分子来不遗余力地制造尽可能多的ATP。食物氧化所释放的能量中，约40%保存到了ATP分子中。

ATP之所以富含能量，是因为它有点像弹簧。它包含三个磷酸基团，像火车车厢那样连接起来。每个磷酸基团都带有一个负电荷，这也就意味着它们之间会相互排斥。但它们又被化学键结合在一起，因而无法逃出彼此的魔爪。当它们被拉开时，磷酸基团的斥力就会突然爆发。[②]

磷酸基团间的化学键可以在与水的反应中被切断，因此这

① ATP并非细胞中唯一的能量源，却是最普遍的。某些酶催化反应会使用其他类似的富能分子，特别是三磷酸鸟苷（GTP）。——原注

② 生物化学的正统派会强调，这只是ATP水解为何释放能量的一种简化版解释而已。——原注

种反应也称作水解。每当有一个键水解，能量就会释放。ATP把最外层的磷酸释放出去，转变成为二磷酸腺苷（ADP）；再将第二个磷酸砍掉就变成单磷酸腺苷（AMP）。两次切断都会释

放出相当多的能量。

消化系统

自食物刚进入口腔起，分解食物的工作就开始了。唾液中包含一种称作淀粉酶的消化酶，能够将糖类聚合物斩断成葡萄糖。这也就是为什么食物咀嚼之后会变得更甜。在蔬菜中，可消化的糖类大多是淀粉。植物细胞壁中的纤维素也是一种葡萄糖的聚合物，但它能抵御淀粉酶的攻击。

在胃里，食物会遭到更严苛的处置。由于盐酸的存在，胃液像电池酸液一样具有了腐蚀性。酸液使得食物分子中的蛋白质螺旋变得松散，以便胃液中的胃蛋白酶来分解。

不过胃还仅仅是个存储未消化食物的器官而已。胃容物离开胃进入小肠后，降解过程更加活跃。小肠液中包含许多种体积小、干劲大的酶，每种酶都专门负责一种拆解任务。淀粉酶分解糖类，其他酶分解蛋白质、脂肪和核酸。分解后的碎片会通过肠道内膜被身体吸收，进入血管和淋巴管，之后营养物质就被输送到身体各处。肠道内膜上有很多微小的褶皱，还有指头状的突起，称作小肠绒毛；这些结构大大增加了肠道的表面积，从而确保营养得到高效的吸收。如果将它们拉平，那么一个人的小肠内膜表面积足以覆盖一个网球场。

很多消化酶都是在胰腺中合成的，胰腺有连接到小肠的导管。这些消化酶分子制造出来是为了分解掉我们细胞的那些组

成成分，那么它们为什么不会破坏我们自身的组织呢？

这些酶在制造时都附带了一个分子保险栓，这个保险栓能够使酶失去活性。这时的酶称作酶原。直到它们到达肠道或者胃中，保险栓（一种肽链的环）才被去掉，一般是被另一种专门负责此项任务的酶去除的。消化道中覆盖有一层黏液，它可以保护消化道不被消化酶分解。如果保护性黏液太薄，那么腐蚀性的消化液就会作用于暴露出的组织上，造成溃疡。

胃容物在进食后数小时内会被分解掉。但即便在消化结束以后，身体依然需要燃料。所以我们的身体要将储备物积蓄起来，以备之后使用。我们肝脏中约十分之一的重量，以及肌肉中约百分之一的重量是一种称作糖原的物质，这是种紧凑、支链很多的葡萄糖聚合物。肝脏和肌肉细胞用收到的部分糖类制造糖原，并制成小颗粒的形状，每个大小约为一毫米的千分之一到千分之四，它们就成了细胞的食品贮藏室。

当血液中的糖含量降低到一定水平以下，身体就知道是时候享用一下存储起来的糖原了。糖含量低会触发胰腺中两种激素的合成——胰高血糖素和肾上腺素，它们告诉细胞开始将糖原分解成葡萄糖。

血液中糖分过多则会触发另一个警报系统，其中就有胰腺中胰岛素的合成。胰岛素能向细胞发出信号，让它们开始将葡萄糖转化为糖原——存储糖类而不是消耗糖类。胰岛素作为糖类供大于求的指标，也能指示细胞去制造别的物质——比如蛋白质和脂肪（另一种应急的能量储备），而停止制造能量。

胰岛素是一种多肽（类似于蛋白质的）激素，由基因编码。有一种常见的遗传缺陷，会导致胰岛素的一种前驱分子（称作胰

78 岛素原）无法正常转化为胰岛素。这种制造胰岛素的障碍正是糖尿病的一个主要原因，也意味着这种疾病的患者需要经常补充一定剂量的激素，以调节血糖含量水平。

齿轮嵌齿轮

糖的燃烧分两阶段，第一阶段过程称作糖酵解（把糖分解），将葡萄糖转化为称作丙酮酸的分子。这一阶段又可以按顺序分解为十种酶分别催化的十小步。分解糖的前五步是上升的过程，需要消耗ATP分子来推动：每分解一个葡萄糖分子就需要将两个ATP分子转化为ADP。不过之后把碎片转化为丙酮酸的五步则是下降的，它可以让ADP重新结合成ATP。这一步中得到四个ATP分子，因此在这一阶段中每消耗一个葡萄糖分子，净得两个ATP分子。所以糖酵解能为细胞储蓄能量。

丙酮酸一般接下来就会进入燃烧过程的第二阶段：柠檬酸循环，这一阶段需要氧的帮助。但如果氧气分子不足，即处在厌氧条件下，应急预案就会启动，丙酮酸会进而转化为乳酸。[①]例如当我们剧烈运动、供氧速率赶不上糖酵解速率时，身体又需要快速满足高能量的需求，这种情况就会发生。当乳酸在肌肉组织中积累时，就会产生酸痛的症状。

作为收割葡萄糖能量的方式，厌氧代谢的效率是比较低的。所以非常剧烈的运动会使肌肉很快疲惫不堪，原因是能量消耗的速率大于能量产生的速率，无论有多少可用的葡萄糖都无济

79 于事。短跑运动员能力有限的原因也在于此。而长跑运动员则

① 我们的身体非常复杂，很少能用普适性的论断来概括，这里自然也一样。有些组织会将大量的葡萄糖转化为乳酸，即便在有氧的条件下依然如此。——原注

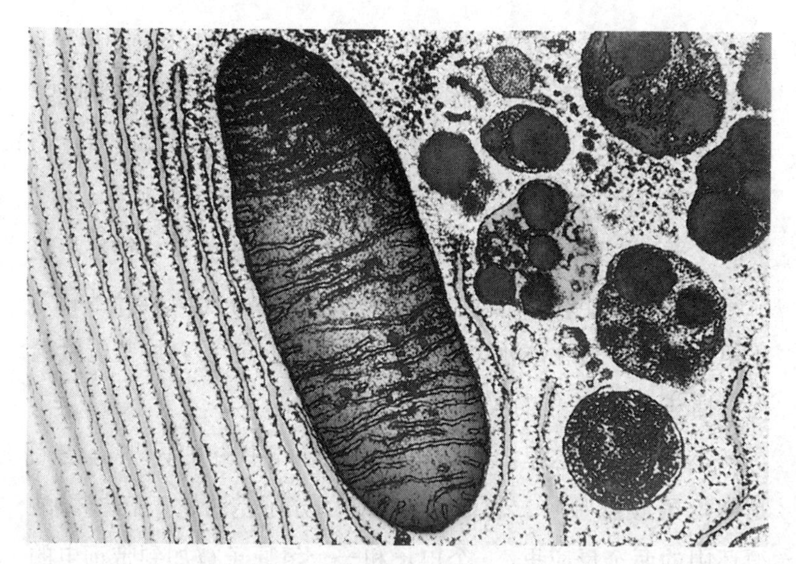

图21　线粒体是细胞中独立的隔间。它们负责生产能量

找到了可以持续的步调，让柠檬酸循环运转起来，有氧（即氧气驱动的）代谢的整个过程全部发挥作用。[①]

这个过程是在线粒体中进行的。这是细胞中一种香肠状的隔间，每个人体细胞里都分布着好几百个线粒体（如图21）。线粒体做的第一件事情是在酶催化下，将丙酮酸转化为一种称作乙酰辅酶A的分子。脂肪分解得到脂肪酸和甘油，最终也会生成这种乙酰辅酶A。

之后的循环是个八种酶催化的反应过程，它先把乙酰辅酶A转化成柠檬酸，接着再转化成其他各种分子，最终得到草酰乙酸这种分子。这个终结又是新循环的开端，草酰乙酸会再次和

① 代谢率更高的动物制造ATP的速度更快，因而更能够经受长时间的运动。蜂鸟能够近乎永远不停地急速拍打翅膀，就像永远不会疲劳的短跑选手。——原注

80

乙酰辅酶A发生反应产生柠檬酸。在循环的某些步骤中，会生成副产物二氧化碳。二氧化碳溶解在血液中，由血液带到肺部，最后呼出体外。因此，事实上原始葡萄糖分子中的碳原子被扔到了最终的产物二氧化碳里面，完成整个的燃烧过程。

不严谨地讲，在循环中被扔掉的还有电子，柠檬酸循环将电流输送到线粒体的另一部分中。这些电子用来把氧气分子和带正电的氢离子转化成水，这是个能量释放的过程。这些能量继而被捕获，制造大量的ATP。

不过电子在这里的流动和在金属导线中的并不一样，它们是由烟酰胺腺嘌呤二核苷酸（NAD）分子携带的。在柠檬酸循环中的两个反应里，一个电子和一个氢原子被加到带正电的NAD离子上，我们把这个新的分子记作NADH。它就是电子的交通工具。NADH分子把电子传递给氧分子，发动氧和氢形成水，同时重新生成NAD。NAD分子又再次投入柠檬酸循环中（如图22）。

大齿轮上还嵌有小齿轮。NADH并非直接把电子给予氧气。这一下降过程又可以分为几步，每一步都给线粒体蓄能。线粒体是由一层外膜围起来的，外膜渗透性较强，柠檬酸循环的主要燃料组分就是通过外膜进入的。外膜里面又有一层内膜，内膜无法渗透。高度卷曲的内膜上布满了酶分子。在内膜上，NADH将电子丢给一个膜上的酶，电子又接龙似的传给其他蛋白质。最终电子到达一种称作细胞色素C氧化酶的膜蛋白质，它的分子上有一处结合氧气分子的位点。电子最后是由这种酶

81

输送给氧的。

细胞色素C氧化酶的氧分子结合位点也可以结合其他比氧

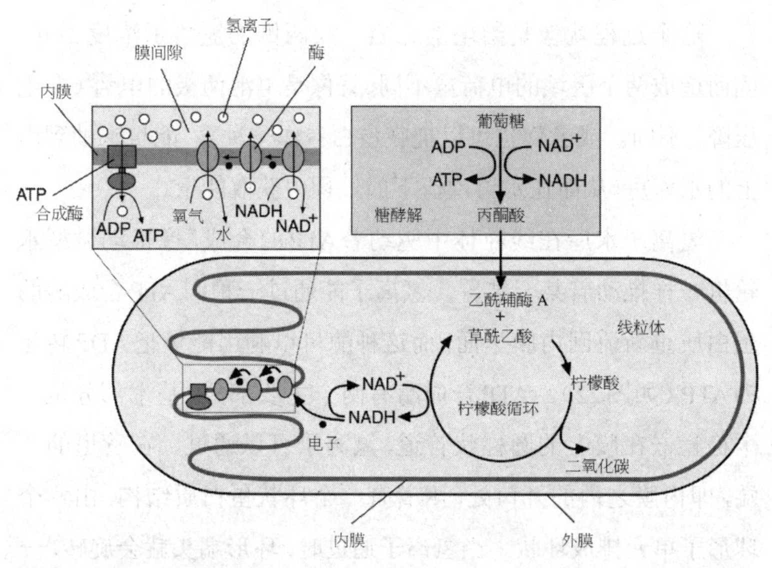

图22　细胞中的能量制造。糖类的燃烧分为两个阶段进行：糖酵解和柠檬酸循环。第一步是线性的反应序列；第二步在线粒体中进行，可以看作紧密配合的一系列循环过程

更强的分子或离子，比如氰化物和一氧化碳。一旦这种情况发生，电子就再也无法传给氧，这部分机制就停止运作。这一层齿轮的阻塞会导致整个线粒体机制陷入停顿，柠檬酸循环终止，整个系统无法再制造ATP。而现存的ATP会在数分钟内耗尽。因此氰化物和一氧化碳都是潜在的有毒物质，可能导致迅速的窒息死亡。任何干扰电子从NADH转移到细胞色素C氧化酶的物质也会产生相似的效果。一些最致命的毒药都属于这类物质。细胞产生能量的这部分机制对攻击非常敏感。

电子传递链中的膜蛋白依次交出自己得来的礼物，释放一部分能量，把内膜内的氢离子推到内膜之外。这样一来，随着电子依次传递，内外膜间隙中的氢离子浓度也就得到了提升。

82

这个过程就像是给电池充电。膜两侧的氢离子浓度不同，因而造成两个区域的电荷量不同，就像是电池两极的电势（即电压降）不同。或者你也可以把膜蛋白看作一种泵，能将水抽到山上的水库里，从而在水再次流下的过程中获取能量。

氢离子水库在线粒体中驱动着ATP的合成，像推动微型水轮机一样推动着某个装置。氢离子能通过一种叫ATP合成酶的蛋白质重新流回内膜里面，而这种酶可以利用能量把ADP转化为ATP（如图22）。ATP合成酶有两个主要部分，基本部分是一个稳稳嵌在膜上的圆柱状管道，氢离子可以通过。在管道的一端，即内膜之内的开口处，附着有一个环状蛋白质结构，由六个球形子单元排成环状。当氢离子通过时，环形端头就会旋转，一旋转就能多造些ATP。ATP合成酶处于细胞产能机制的中心，1999年的诺贝尔化学奖颁发给了保罗·博耶、约翰·沃克和延斯·斯科，他们的贡献正是解释了ATP的结构及许多作用机理。

因此，糖酵解和柠檬酸循环是非常不同的两种代谢过程。一个是无氧的，一个是需氧的。糖酵解有可能自封，虽然其中一步需要用到常在柠檬酸循环中产生的NAD，但NAD也可以通过将丙酮酸转化为乳酸在厌氧条件下重新生成。糖酵解和柠檬酸循环这两个过程很像是两种独立的代谢途径绑定在了一起。

而这很可能恰恰正是它们的情况。人们认为，线粒体曾经是分立的细菌生物体。有证据能支持这种说法：它们拥有自己的DNA，与细胞核中的主基因库完全不同（参见第45页）。人们相信约20亿年前，线粒体开始了与利用糖酵解途径进行无氧呼吸的单细胞生物的共生关系。

大约在那时，绿藻（原始的植物生命）的蔓延导致地球大气中氧气含量的急剧增加。而在之前，空气中的氧非常少。那种类线粒体的细菌能够利用柠檬酸循环“呼吸”掉氧气。无氧细胞制造丙酮酸，需氧菌能用作燃料，同时需氧菌制造NAD，能够帮助无氧细胞的糖酵解过程，因此两者间的共生关系能够发展起来。最终，无氧细胞吞噬了需氧菌，两者合为一体，组成复合的有氧呼吸细胞。后来的就众所周知了。

酵母菌则从来没经历过这些，它仍是糖酵解厌氧菌，靠糖类生存。不过酵母细胞并非将丙酮酸转化为乳酸，而是转化为酒精，千百年来为我们带来啤酒的香气。这个过程就是发酵过程，历史上正是从这个过程出发，我们才日渐理解了酶的催化作用。发酵过程中也会产生燃烧的最终产物二氧化碳，它们会在世界每个角落的发酵缸中冒出泡泡。

结合氧气

氧气通过肺部进入体内，但它们并不容易溶解在血液中，因此也就无法简单通过溶解的办法传递到线粒体处。（相反，二氧化碳则易溶，于是就从细胞跟随血流到达肺部。）氧气会被血红蛋白这种蛋白质打包，它对氧气分子的亲和力非常强，打包起来由血红细胞输送。血红细胞几乎完全是为了这一个任务而存在的。它和别的细胞不一样，基本上就是一包血红蛋白，不包含DNA和RNA，细胞里没有各种隔间，几乎再无其他酶。细胞成熟以后，制造血红蛋白所需的全部遗传和酶催化部件会在最后时刻被排出。

血液的红色与铁锈的标志性红色很相似。血红蛋白分子的中心是铁原子，它被圆环状的卟啉分子团团包围。卟啉分子加

84

上铁原子即为一个血红素基团，它们吸收蓝光和绿光的能力很强，因此看上去呈鲜艳的红色。每个血红蛋白分子包含四个血红素基团，其中铁原子的位置为氧分子提供了附着位点。

作为氧气的输送者，血红蛋白必须牢牢地束缚住它的货物，同时也得让货物能够脱离。怎样才能两全其美呢？血红蛋白用了一种聪明的诡计，蛋白质常常都会用这种小诡计，即结合目标分子后就改变自己的行为。简单地讲，就是蛋白质的一部分能够“感觉到”另一部分发生了什么。

肺部血管中有很多氧气分子，当一个血红素基团附着上氧分子后，会发出一个震动信号，促使另外三个血红素基团也去捕捉氧分子。血红蛋白基本上是在肌肉组织中给出氧分子，传递给肌红蛋白。肌红蛋白是另一种结合氧分子的蛋白质，对氧的亲和力比血红蛋白还要强。当一个氧分子离开血红蛋白时，“反震动”就会削弱其他三个血红素基团结合货物的能力，于是它们也就欣然释放了氧分子。这些震动称作变构，震动虽小，却是蛋白质链折叠构型的显著变化。

所有脊椎动物和不少无脊椎动物都用血红蛋白来运送氧分子，不过不同的物种其确切的蛋白质形状有所不同。节肢动物和软体动物则使用另一种结合氧分子的蛋白质，称作蚯蚓血红蛋白。这种蛋白质里也有铁原子，不过不是被血红素基团束缚住的。它结合氧后会呈现介于紫色和粉色之间的颜色，未结合氧时则无色。有些海洋无脊椎动物会用另一种金属元素来输送氧，它们用来结合氧的蛋白质称作血蓝蛋白，呈现蓝色，也就说明其中包含铜元素。它们可谓海洋中名副其实的“蓝血贵族”。

绿叶的能量

地球大气之所以富含氧气，靠的是植物的功劳。光合作用是生物利用光的能量来制造分子，而氧气就是光合作用的副产品。因为植物和光合细菌位于食物链的最底层，所以地球上全部生命的能量最终都来自太阳能。没有植物我们就都完了，不但我们自己每天没有面包吃，牛羊也只能在不长草的贫瘠牧场中饿死。

光合作用是一种非常非常古老的生命过程。至少35亿年前，大陆才刚刚形成、寸草不生的时候，藻类就开始进行光合作用了。这些生命体是第一批**自养生物**——自己养活自己，能够用光、水和空气中的碳等原材料来制造自己的分子。每年大约有600亿吨的碳会从大气中的二氧化碳被转化为富含能量的生物质。我们把其中的一部分吃掉，烧掉，盖房子，造桌子，喂给牲畜，浆成纸张，织成布匹。而更多的能量则落到地上，被微生物分解，变成易挥发的含碳化合物重新回到大气中。经过地质学尺度上的很长时间，很多会埋在地下，压缩成为煤炭，或是分解成为石油或天然气。

光合作用依赖于能够与光相互作用的分子，分子需要吸收光的能量，并把能量传递到化学过程中。如果说线粒体是哺乳动物细胞中的锅炉房，那么叶绿体就是植物叶子细胞里的光能中心（如图23）。大致说来，叶绿体里所发生的反应过程正是葡萄糖代谢的逆过程，它利用二氧化碳和水制造糖类。“燃烧”葡萄糖是个能量下降的过程，于是我们就知道，光合作用中产生葡萄糖就是个能量上升的过程。正因如此，植物才需要光的能量

86

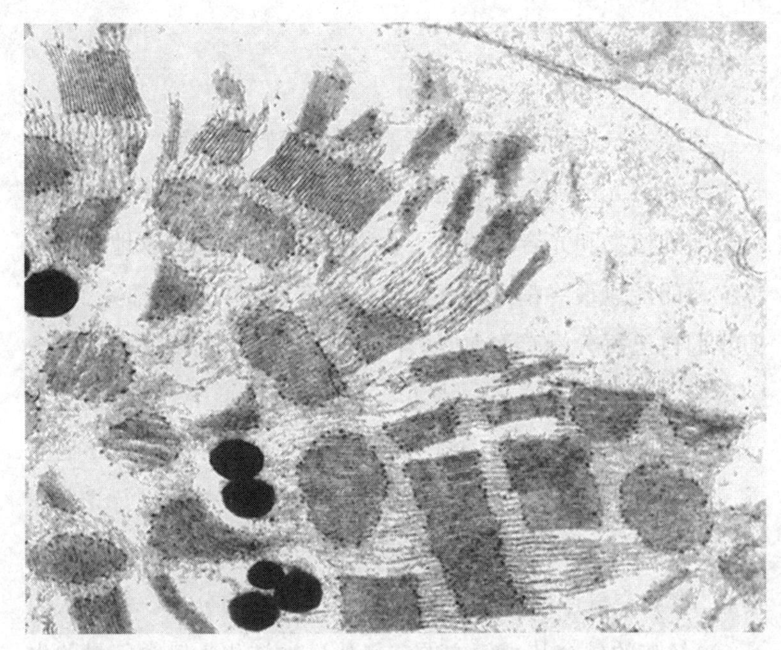

图23 植物在叶绿体中捕获光能，并将其转化为化学能。叶绿体中堆积着一层层的类囊体膜

参与进来。植物利用这些能量并不仅是制造葡萄糖、然后编织进细胞壁里的纤维素而已，与之同等重要的还有用能量来产生ATP分子，驱动细胞里的化学反应。

有氧代谢和光合作用有若干相似之处。两者都包含两个独立的子过程——线性的反应序列与循环过程耦合，循环过程产生两个子过程都需要的分子，而两个子过程有着不同的进化起源。糖酵解和柠檬酸循环间的桥梁是作为电子摆渡船的NAD分子，而光合作用中两个子过程间的桥梁是个几乎一模一样的分子——磷酸NAD（NADP）。

87

在光合作用的第一阶段，光用来转化NADP，让它携带上电子（即转化成NADPH），并将ADP转化为ATP。这实际上是一个储能的过程，为叶绿体合成葡萄糖打下基础。第二阶段是个循环过程，称为卡尔文–本森循环，在这里ATP和NADPH用来将二氧化碳转化为糖（如图24）。

叶绿体中有层层叠叠的膜，称作类囊体膜，第一阶段就在膜表面上发生。类囊体膜上散布着一簇一簇的分子，称作“光系统”。在光系统中，能够吸收光的分子称作光合色素，它们开启了光能驱动的反应。光系统的核心称作光合反应中心，位于这个中心的是叶绿素a分子。叶绿素a能够大量吸收红光和蓝光，因而使叶子看起来呈绿色。

88

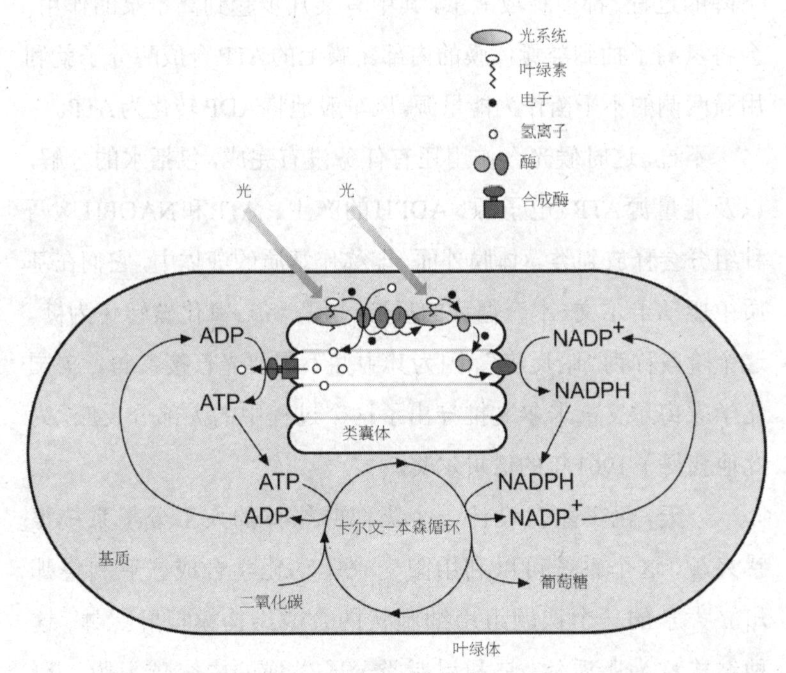

图24　在光合作用中，捕获的光能用来分解水、制造ATP，两者继而又驱动二氧化碳向葡萄糖转化

当叶绿素得到光能时，它会被“激发”，就像苹果树被摇晃。在激发态上，叶绿素束缚外层电子的能力减弱，于是其中一个电子就脱离它成为自由之身。这个电子接着会传递给一个酶。当酶收到两个从叶绿素“摇下来”的电子时，它就能把一个NADP阳离子转化成NADPH。在另一个光能驱动的反应中，缺了电子的叶绿素能重新从水分子中夺得一个电子。而水分子被分解为氢离子和氧原子。氧原子两两结合形成氧气分子，植物通过叶子表面的气孔把氧气释放出去。

叶绿素从水中得到电子的过程，实际上经过了一连串的分子传递，它们都嵌在类囊体膜上。传递过程的每一步都是能量下降的过程，都会释放能量，其中有某几步起到离子泵的作用，会将氢离子抽到类囊体膜的内部。膜上的ATP合成酶分子就利用膜两侧的不平衡作为能量源，风车般地将ADP转化为ATP。

不过，这时候光合作用还有任务没有完成，包括水的分解，以及能量源ATP、电子源NADPH的产生。ATP和NADPH这两种组分会释放到类囊体膜外面，即称作基质的液体中，它们在基质中驱动卡尔文-本森循环的反应过程，将二氧化碳转化为糖。这个阶段称为“暗反应”，因为其中并不需要光直接参与。美国化学家梅尔文·卡尔文推导出了这个过程中的大部分反应，为此他获得了1961年的诺贝尔奖。

89

今天，化学家对设计一个类似叶绿体的人工分子系统很感兴趣，这个系统可以利用阳光来驱动化学合成。亚利桑那州立大学的一个课题组用细胞状的合成结构模拟叶绿体，这种结构称为脂质体，是利用脂类分子做成的中空球形膜。研究人员在类脂体膜上撒有一些设计好的分子组合体，使它们能够执行与光合反应中心相同的任务：利用光能将氢离子抽入脂质体的内部。

研究人员将ATP合成酶分子注入他们的脂质体里，它们可以释放氢离子，并顺带制造了ATP。人们寄希望于ATP存储的能量能够用来进行化学合成，比如去执行某些工业上有用处的生化反应。

轰然结尾

葡萄糖和蜡（一种烃类）都“蕴含能量”，利用氧气断裂它们的化学键，就可以以热能的形式释放出能量（除非将能量导入其他形式）。但是，有些化学家却还在寻找蕴含着更多能量的分子。那么一个分子里究竟能装得下多少能量呢？

阿尔弗雷德·诺贝尔就在19世纪为回答这个问题而奔忙，结果却是一个广为人知的反讽：他发明炸药积累了大量财富，却用财产资助了一项著名的年度和平奖金。诺贝尔的创新，并不在于他发明了一种富含能量的分子，而在于他找到了一种方法把已有的炸药包装成另一种形式，减少了它趁人不备就随意爆炸的机会。

最古老的炸药是黑火药，它是硫黄、硝石（硝酸钾）和炭的混合物，约公元11世纪在中国发明。在西方，人们未加多少改动就直接应用它（造成了很糟的后果），直到19世纪科学家开始寻求制造更猛烈爆炸的办法。1845年，瑞士化学家克里斯蒂安·舍恩拜因发现了硝化纤维，即用硝酸和硫酸对棉纤维（纤维素）进行处理得到的化合物。这是第一例“半合成”聚合物，是自然界和化学家手艺共同的创造。人们后来又用硝化纤维来制造赛璐珞——一种硬塑料，还有第一例人造丝。但硝化纤维也具有爆炸性，所以就成为了著名的火药棉。

火药棉暴烈的脾气很难控制：人们最初尝试批量生产它时，造成了好几例死亡事故。1847年，意大利化学家阿斯卡尼奥·索布雷罗合成了一种相似的危险物质，称作硝化甘油。阿尔弗雷德·诺贝尔在1859年开始研究这种化合物，试图找到一种办法使它变得稳定，除非人们刻意让它爆炸。虽然1864年的一场爆炸事故炸死了他的弟弟，诺贝尔依然坚持研究，最终发现，将一种称作硅藻土的黏土与硝化甘油混合能够得到泥灰状的炸药，可以让人们安全地操控。他给这种炸药起名叫甘油炸药。1875年，诺贝尔发明了炸胶，这是硝化甘油和火药棉的胶状混合物，比两种物质各自分开的威力都要大。

这些炸药大部分用于采矿和建设中的爆破，这给诺贝尔带来了财富。但它们终归无可避免地也会用于军事。19世纪末，英国和法国军队使用了无烟火药——一种类似于炸胶的炸药。德国军队使用了三硝基甲苯（TNT），只有用一种次级的炸药引爆它才会爆炸。阿道斯·赫胥黎在《美妙的新世界》中向我们讲解了这种致命混合物的奇特的化学特性：

> $CH_3C_6H_2(NO_2)_3 + Hg(CNO)_2$[①]等于，啊，什么？等于地上的一个巨大的窟窿，一大堆破砖碎瓦，几片肉和黏膜，一条腿飞到天上叭的一声掉下来，落到天竺葵丛里……[②]

① 加号前的物质即TNT，加号后的化合物雷酸汞就是作者上文提到的用作引爆的物质。——译注

② 译文节选自孙法理译，《美妙的新世界》，译林出版社2000年6月版。——译注

上面所有这些化合物都是包含硝基的有机物质。所谓硝基就是一个氮原子连着两个氧原子。硝基化合物之所以有爆炸性，是由于这些材料在点燃时会使氮原子重新形成氮气分子。氮气分子的化学键非常稳定，在形成过程中释放大量的能量。同时，氧原子能够促进燃烧过程，使得燃烧迅速发生。开发威力更大的炸药，很大程度上就是要想办法把更多的硝基结合到有机化合物中。RDX（或称黑索金）炸药就通过这种办法改进了TNT，如今用在武器当中。目前所生产的威力最大的富氮化合物称作HMX，是高熔点炸药的缩写。

2000年初，芝加哥大学的化学家们设计了一种可能具有更多能量的硝基化合物。这种化合物名叫八硝基立方烷，包含八个碳原子组成的立方体，每个碳原子又连着一个硝基。这个分子不仅饱含硝基，而且立方体形状意味着碳原子间的键张力很大，很容易断裂。另外，由于分子形状紧致，它应该可以堆积起来形成密实的晶体形式。最初的实验还没有实现这种高密度的形式，但是计算结果已经预测，如果能造出这种形式，那么就单位质量来说它的爆炸能量将高于任何一种已知的非原子能炸药。

阿尔弗雷德·诺贝尔将遗产用于颁奖，可能是由于他的成果被用于杀伤和毁灭，他为此感到后悔，从而做出了这样的表示。但是很明显，并非所有化学家在面对与武器相关的研究时都会有这样的情感。在我个人看来，这种工作可以说是科学的不端。可更重要的是，对烈性炸药的研究表明，科学研究无法截然分成“纯粹”的、无关道德的科学部分，和低等的、满足社会“应用”的科技部分。制造八硝基立方烷确实是高超技术取得的辉煌成就，但它也得到了美国国防部的资助。

92 炸药这个例子向我们透露了分子科学的两面性。爆炸其实也是化学的一部分乐趣所在——在学校实验室里就能实现的经典爆炸反应，有几个崭露头角的化学家从没想要做做看呢？我所知道的就至少有一位德高望重的科学家曾经被学校开除，因为他差点把学校毁了。（他现在在研究能灭绝所有生物的巨大陨石的影响。）不过，从这些小爆炸、小闪光到德累斯顿和汉堡[1]的 93 毁灭，只需要一小步。这是凶险的一步。

① 德国城市，二战期间遭到盟军的轰炸。——译注

第五章

运动的精灵：分子马达

餐后演讲一般并不适宜发动革命。但理查德·费曼1959年在美国物理学会西海岸分会上的演讲是个例外，他不是个常规的物理学家。费曼是20世纪战后最具创造性的科学思想者之一，这个世界铭记着他鲜活的形象——邦戈鼓手，恶作剧大王，保险箱破译高手，现代科学中一个搞怪的形象。

费曼1959年的演讲意气风发，但他的意图是相当严肃的。他称演讲的题目为“底下还有充足的空间”，内容是有关肉眼无法看到的微小尺度上的工程。“我所想要讲的，”他说，“是在小尺度上操纵、控制事物的问题。”费曼说，所谓“小”，并不是指“小拇指指甲盖大小的电子马达”。他指的是像原子一样小的尺度。

“想象一下，”他接着讲道，“假设我们能够一个一个地按照我们的意愿来安放原子。”他看到，这实质上正是化学家努力要做到的：

> 当化学家想要造一种分子时，他会做神秘的事情。他

> 发现必须要获取那种环，于是就把这个和那个混合在一起，摇一摇，瞎弄弄。经过一番困难的过程，最后他一般都能成功合成想要的东西。

94

你能看到，在物理学家眼中，化学家并不比外行高明多少。不过，费曼这番描述倒与普里莫·莱维对化学家如何像工程师建造桥梁一样建造分子的说法有些异曲同工之妙（参见第24页）。但其中的区别是，化学家习惯于将分子看作一种物质，能够结晶的东西，能够放在瓶子里的东西。而物理学家则将它看作一种结构体，就像是发动机的零部件。

本质上讲，费曼所思考的其实是物理学家能否找到办法去做化学家的工作，但他披了一件工程师的外衣。我们能不能通过一个一个码放原子来制造分子呢？1959年的时候，这件事对任何人来说都是无法设想的，但想象力魔术师费曼就想到了。

不过他并非毫无根据地瞎想。即便在当时，事实也清楚地表明技术会向越来越小的方向发展。1940年代晶体管的发明缩小了电子学的尺度。人们替换掉放满真空管的笨重箱子，转而使用密集式器件，里面包含了硅晶体管制成的“固态”电路。便携式晶体管收音机出现在了美国的每一处海滩上。当时工程师在制造微小的机械零部件方面的技艺水平已经日益提高——实际上远高于费曼所认识到的程度。费曼当时提供了两笔奖金，各一千美元，希望能够对微型化技术的发展起到些小小的促进作用，奖金由他提供：第一项是制造出各方向尺寸均不大于1/64英寸的电子马达，第二项是将书上一页纸的内容写在微缩的面积上，按比例缩小为原来的1/25 000。费曼大概以为这笔钱好几年都不会有人拿走，可没想到有人（一位名叫威廉·麦克莱伦的工程师）短短几个月后就完成了他的第一项挑战。

今天我们可以走得更远。我们能够用酸蚀法或者电子束法在硅片上雕刻微小的齿轮和马达，大小只有十分之一毫米（如图25）。但若将尺度缩小到大约十分之一微米，要在材料片上蚀刻就不好办了：当前使用的硅集成电路制造方法有局限，蚀刻出的导线只能细到这种尺度。要是比这个尺度还要小，这种方法就不能用了——这就好比要用切面包的刀劈开人的头发丝一样。

95

研究人员开始思考，这种“自上向下”的方法在这种尺度上是否还有意义。其实这种小元件的尺度和分子更加接近（中等大小的分子是这个尺度的百分之一），离咱们看得见摸得着的硅

图25　硅片上刻出的微马达

片反而更远。那我们是否应该从单个的分子出发，自底向上地来制造东西呢？

普里莫·莱维在《猴子的扳手》中承认，化学家们幻想着一种能用于分子尺度建造的工具包：

> 我们晚上做梦都想要一套前所未有的镊子，就像干渴难耐的人梦见一泓清泉。这种镊子能让我们夹起一段分子，牢牢地夹住，让它拉直，然后按正确的方向和已组装好的部分粘在一起。如果能拥有那种镊子（可能某一天会有的），我们就可以创造很多迄今只能由万能的神创造的漂亮东西，组装青蛙或蜻蜓大概还不行，但至少组装个微生物或者霉菌孢子还是可以的。

费曼同样在分子器件和人工生物中找到了灵感：“在其中化学作用力会反复地使用，制造出各种奇怪的效应（也包括作者这种奇怪的东西）。”他意识到生物学中已经有分子机器的存在。在1959年，若是这番演讲被生物学家听到了，必定会被斥为愚蠢的物理学家妄图将自己的个人观点强加到他一无所知的领域。不过今天的生物学家已经很乐于用分子机器的名义来讨论蛋白质。

96

这一章，我们就来看看最引人注目的那些产生运动的蛋白质分子。它们就是分子马达，也常称作马达蛋白。和它们对比起来，当初赢得费曼奖金的那个微马达只是个粗笨的庞然大物，就像拿笨重的恐龙和灵敏的跳蚤相比。马达蛋白的生物重要性无可估量。若是没有这些蛋白质，我们就无法活动肌肉，鸟无法飞上天空，鱼无法横渡海洋，连细菌都无法活动。还有更严重的后果：细胞无法分裂，于是就再也不会有繁殖。失去驱动运动的分子，生命将不复存在。

但就分子世界的工程师而言，马达蛋白还告诉了我们别的东西。它们表明，分子尺度的工程是可能实现的：我们可以把日常世界中熟悉的思想缩小到分子的王国。从这个角度看马达蛋白并非唯一的，只不过它们十分明确地向我们展示了这一点。我将讲述我们有可能怎样通过制作定制的分子马达，从零开始来实现类似的目标。这将引领我们走向**纳米技术**（纳米尺度下的技术，纳米即为可用分子来度量的长度）的竞技场。在这条路上，理查德·费曼的演讲就是第一处清晰的路标。

细胞自由泳

分子的形状永远都不是固定的，它疏松的部分会一直振动、摇摆。分子世界中，机械运动无处不在。

但一般而言，分子运动要么是随机的——如聚合物长链漂在溶液中弯曲着蠕动，要么是平均化之后为零——如化学键伸缩式的振动。而真正的马达与此相反，我们需要它朝特定的方向运动——可以称之为有目的性的运动。

任何马达都要消耗燃料。你可以把这看作有规则运动无可避免的成本，是热力学第二定律强制的代价。而相反，随机分子运动就可以“免费”地进行，它无规则的分子摆动正是热的体现。

我们体内进行着很多种带有方向性的运输。比如，纤毛（像毛发一样附着在呼吸道中，包括肺部和气管）的运动能将一层黏液从肺部一直输送到喉咙，在这里就会聚集成为痰。这种黏液能够捕获灰尘，因而输出它可以维持肺部的清洁。为了将黏液向上送到气管，纤毛不能只是随意摆来摆去，而需要步调协调，有秩序地移动，就像游泳时手臂运动那样。它们运动时先像鞭子一样来一记“猛抽”，之后再缓慢地“回复”。某些单细胞原生
98 生物的确就利用细胞表面的纤毛在水里游动。

驱动这种运动的分子马达是一种蛋白质，称作动力蛋白。每根纤毛都含有微管（参见第63页），微管围成一圈形成一支更大的管，称为轴丝。微管两两一组结合成双管，就像双筒枪那样。每个轴丝有九对双联管，双联管之间通过动力蛋白相互连接，动力蛋白就是双联管上的小突起，每隔一段就有一处，分布得很有规律，像多足的蜈蚣（如图26）。

为了移动轴丝，微管会交错爬行。每个动力蛋白分子都有“一条腿”，消耗ATP发生反应可以使“腿”弯曲。动力蛋白大体
99 上就是一种酶，它分解ATP来改变自己的形状。这个反应同时还需要钙离子来触发。神经信号可以通过控制是否向纤毛注入钙离子来控制这个运动。

由于动力蛋白分子指向的方向都相同，因而微管弯曲时，双联管中的一支就会沿另一支身上爬行。如果分子只是单纯地又再次伸直，那么微管也就会回到原来的位置。为了产生向前的运动，每个动力蛋白在伸直之前，会先将自己从另一支微管身上解离开来，伸直后又再次附着在另一支微管身上，为下一轮“动力冲程”[①]做好准备。只有当动力蛋白的“脚”附着在另一支微管身上时，它才能够分解ATP并转换到弯曲状态。

① 燃气机在循环中点火耗能、产生运动的关键一步。——译注

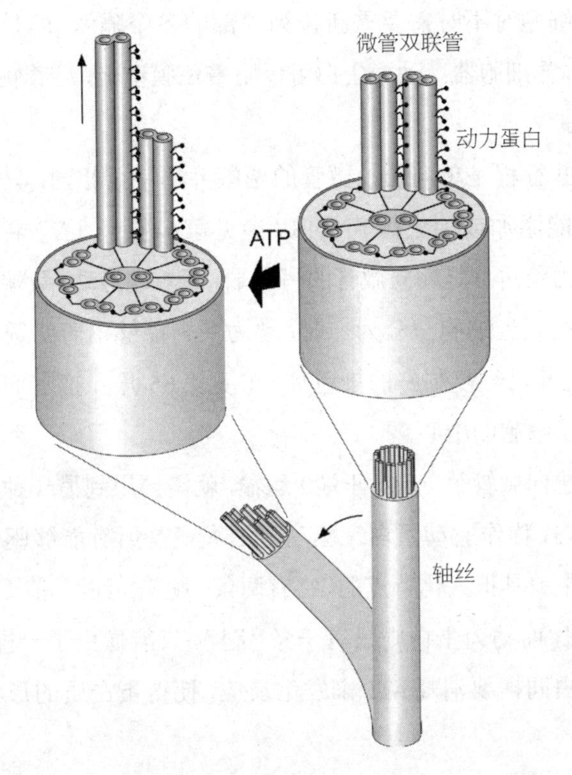

图26　动力蛋白分子马达驱动纤毛的弯曲

所以，双联管的交错运动过程就类似于棘轮，不断重复着附着、弯曲、解离、伸直的循环过程，动力蛋白就在这个循环中产生了向一个方向的运动。而由于微管的末端固定于轴丝的底部，这样的交错滑动就会带动鞭毛弯曲。滑动若能够协同配合，鞭毛就能够来回弯曲。这个协作似乎是由于一对微管沿着轴丝中心运动，但具体怎样实现的人们还不清楚。

动力蛋白在细胞世界中发挥的作用更加普遍：它就是负责输送货物的引擎之一。我们的细胞中交织着微管构成的内部铁路网。细胞时不时需要重新排列内部的各个隔间，即带有膜的结构，称作细胞器。动力蛋白可以附着在膜壁上，沿着轨道推动细胞器。

这些行程是单向的。微管的两端并非完全相同：只有其中一端才能添加或除去微管蛋白（参见第64页），这一端称作正端。动力蛋白总是朝向微管的另一端，即负端运动，负端位于细胞的中心。当细胞一分为二时，动力蛋白能够沿着纺锤体中的微管，拉开已经复制好的染色体（参见第65页），把它们拉向两个新生子细胞的中心。

100

要想向微管另一端（正端）运输，就需要用到另一种不同的马达蛋白，称作驱动蛋白。驱动蛋白大概是所有能够驱使运动的分子中最具拟人化特点的，它有两条“腿”，可以一前一后“蹒跚而行”，而动力蛋白则只有一条“腿”，只能像虫子一样蠕动。驱动蛋白同样要消耗ATP来发生反应，使得蛋白质的形状发生变化。

驱动蛋白是细胞的邮递员，将包裹从一个细胞器运送到另一个细胞器。比如，蛋白组装好之后，需要从它的制造地（内质网）运送到细胞中需要它的地方。于是它们会打包进一个膜结构的小球之内，称为运输囊泡，然后由驱动蛋白携带着运输囊泡沿着微管网络运送到合适的地方。

肌肉的力量

我们人类自己的行走是通过肌肉的收缩和拉伸而实现的。骨骼肌一缩一伸，像弹簧一样收紧舒张，就控制了我们所有的运动，从钢琴家手指灵巧的飞舞到运动员大腿健壮的弹跳。

骨骼肌也是大自然具有层级结构的一种分子材料（参见第53页）。它由一束一束的纤维组成，纤维又由小纤维组成，小纤维又由更小的纤维组成。每一个肌肉细胞都非常细长，里面包含很多股由纤维编织而成的绳索，称作肌原纤维。这些绳索拥有复杂的分子亚结构，肌肉收缩的秘密就蕴含在其中。

在显微镜下观察，肌原纤维有不同宽度的明暗条纹。骨骼肌在高放大倍率下呈现条纹状，因此我们也称其为横纹肌。条纹序列周期性地沿着肌原纤维重复出现，一个重复单元称为一 101 个肌节。肌节中不同的区域还拥有通俗的名字，如A带、H带等——人们之所以这么叫恰恰也暗示了，当它们最早被观察到时，谁也不知道它们到底表示什么。

安德鲁·赫胥黎、休·赫胥黎和他们的合作者在1950年代注意到，当肌肉收缩时，这些带状区域的宽度会变化，于是他们提出了肌肉运动的肌丝滑行理论。他们认为，肌原纤维包含牙刷头形状的结构，且这样的结构两两相对，于是彼此的刷毛相互交叉。每个肌节含有两组这样的刷头组结构，处于背对背的位置。暗带就是刷毛交叉的地方（从而分子密度较大），而亮带则是只有一侧刷毛的地方（如图27）。两位赫胥黎提出，肌节中的刷毛可以相互交叉得更深，于是肌原纤维缩短，使得肌肉收缩。 102

肌丝的相对运动是肌球蛋白这种马达蛋白所驱动的。肌球蛋白呈瘦长形，上面有两条螺旋链相互盘绕。链的两端终止于梨形的端头。肌球蛋白分子聚集成束，称作肌球蛋白丝。肌球蛋白丝刷毛的每个末端都布满肌球蛋白的端头，像扎成一捆的玉米秸秆，顶端有一颗颗玉米。

穿插在肌球蛋白束中间的是由另一种称作肌动蛋白的蛋白

质组成的肌丝。这种蛋白质事实上是球形的，小球会连接成链状，像项链上的珠子。肌动蛋白丝上，两条肌动蛋白“珠链”彼此盘绕，又形成一种双螺旋。项链上还装饰有原肌球蛋白形成的细线，沿着肌动蛋白丝一起盘绕。项链上每隔一段距离还有一个球形的肌钙蛋白（如图27）。

当肌球蛋白的端头将自己附着在肌动蛋白丝上并拉动时，肌肉就发生了收缩。它的原理与动力蛋白、驱动蛋白沿微管运动的原理是一样的：ATP分解为ADP，推动附着的马达蛋白形状改变，从而产生运动。肌球蛋白头部通过铰链似的颈部与分子其他部分连接在一起，头部可以围绕颈部摆动。头部发生一连串动作：旋扭，从肌动蛋白上解离，解扭，重新附着到肌动蛋白上，循环往复，间歇性地发动动力冲程，如棘轮般沿着肌动蛋白丝转动（如图28）。

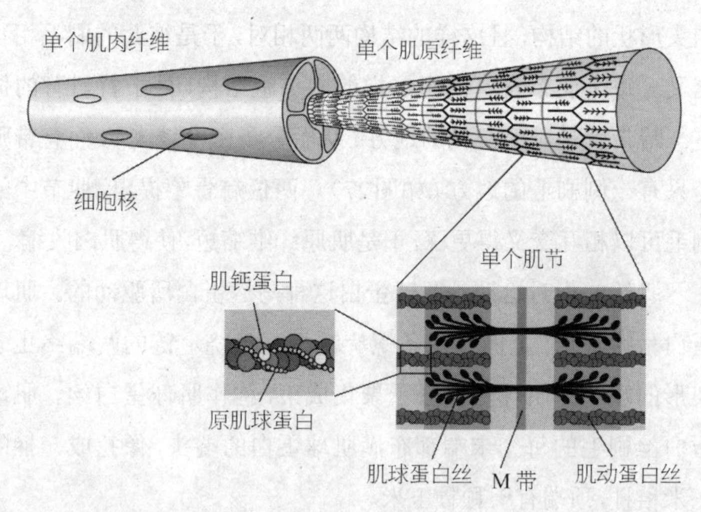

图27　肌丝的交叉结构使得肌肉能够收缩

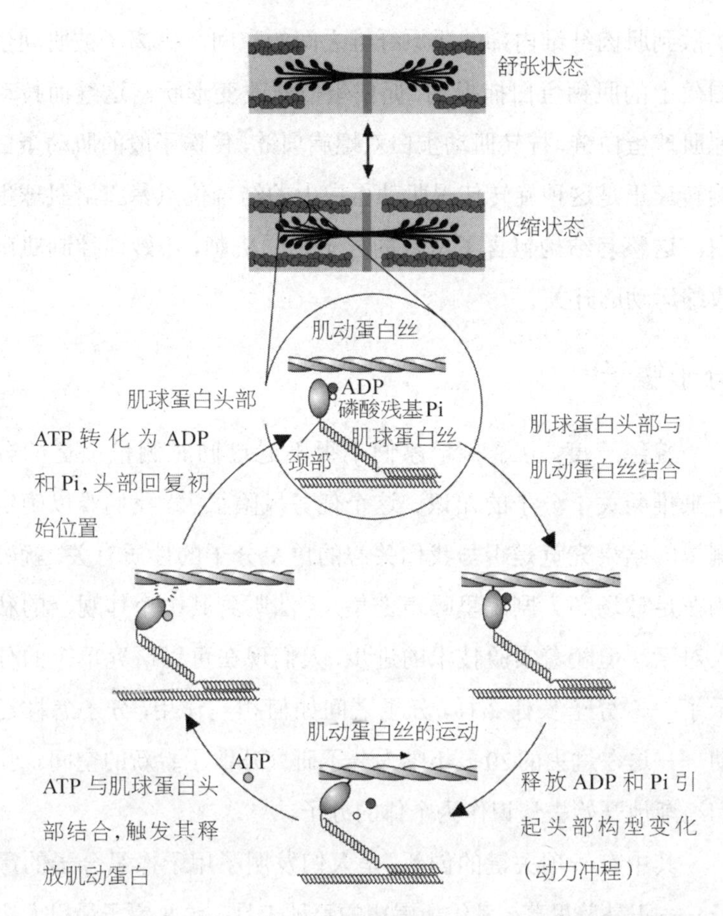

图28　肌球蛋白分子马达沿着肌动蛋白丝行进,引发了肌肉的收缩运动

这整个过程都受到自发的控制，从大脑处来的神经冲动告知肌肉是收紧还是放松，从而产生相应的运动。肌动蛋白丝上的原肌球蛋白和肌钙蛋白就起到这个开关的作用。肌肉通过称为运动神经元的神经细胞与大脑相连，像一种生物化学式的电线“焊接”在肌肉纤维外部。电信号传递到运动神经元末端时，就会触发钙离子从一个管道网络——肌浆网——中释放，并

103 扩散到肌肉纤维内部的肌原纤维之间的空间。钙离子被肌动蛋白丝上的肌钙蛋白捕捉到，促使蛋白质改变形状。这继而拉动原肌球蛋白链，拧转肌动蛋白双螺旋项链，使珠子般的肌动蛋白旋转。正是这种旋转使得肌动蛋白上的结合位点暴露给肌球蛋白。这整套结构包含了精致的分子传递机制，巧妙地控制肌肉 104 收缩运动的开关。

分子镊子

曾经一度，分子科学家们不得不通过同时测量上亿的分子来推断关于分子的知识。这个任务风险很大，我们难以确定测量的结果究竟是否与我们关心的单个分子的性质有关，就好比在足球场和大剧院里噪声密集，无法听到其中个体观众的私人对话。但随着实验技术的进步，人们现在可以研究单个的分子了——分子长什么样，分子之间如何相互作用，分子怎样运动——这在过去的20多年中为分子研究开辟了崭新的空间。我们渐渐地开始去认识作为个体的分子。

其中有一项关键的创新，是人们发明了用于操纵分子的镊子——正是普里莫·莱维所渴望的那种工具。这种镊子最引人注目的并非它有多精致，而是它完全无形，是由光构成的。它称作光镊，利用高强度的激光束来捕捉目标。我们日常就机械马达会关心一些基本问题：效率如何？负荷量有多大？跑得有多快？而光镊使得研究者们能够回答这些关于马达蛋白的类似问题。

光与分子中的电子之间相互作用，能够在目标物质上产生一种力——一种“光压”。若目标足够小，光的强度足够大，那么这种力就足以移动目标。当两束乃至多束激光交叉时，会在光镊中产生一个极亮的光点。这个亮区中的小目标会受到来自各个方向的光压，阻碍它向各个方向运动。于是，它就被激光束镊子构成的陷阱给抓住了。当光束移动时，目标物也会随之被拉动。

于是我们可以测量单个马达蛋白所产生的力，只要将马达蛋白（或者它所沿之运动的物质，如肌动蛋白丝）拴在被光镊夹住的微观小塑料球上。马达产生的运动能够将小球拖出陷阱中心，而位移量就正比于马达产生的力。

105

利用小球作为操纵把手，就可以用光镊对分子做非常特别的事情。日本庆应义塾大学的木下一彦和同事们将小珠附着于肌动蛋白丝的两个末端，然后拉动两端使它们穿过自身形成的环，创造了一个分子纽结（如图29）。它们将纽结拉紧直至断开。由

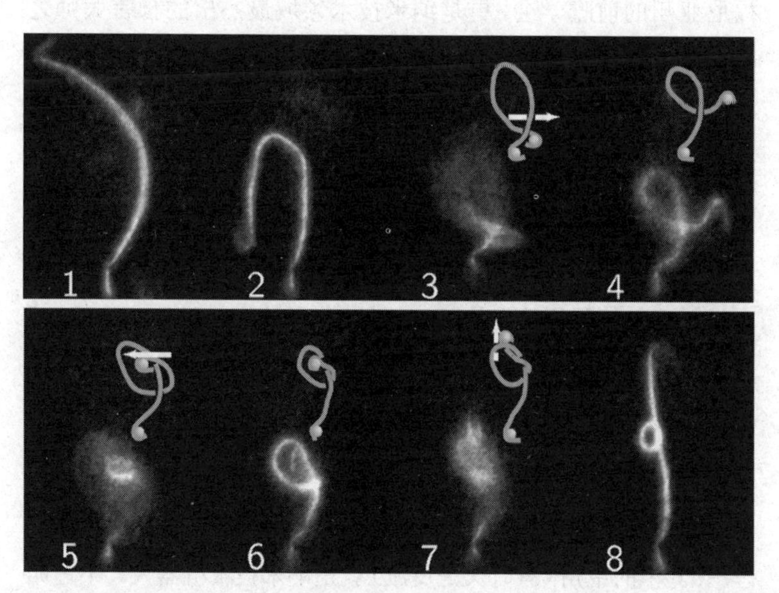

图29　光镊可以用于将一股肌动蛋白丝打结。附着于肌丝两端的微观小球充当着“操纵把手”。用光线照射肌动蛋白使其发出荧光，从而在显微镜下可见

于肌动蛋白丝较为僵硬，如同树苗的枝杈，因此在高度弯曲时就尤为脆弱，拉断打结的肌丝就远比拉断未打结的所需力量小。

106

光镊并非操控单个分子的唯一工具。实践证明，1980年代设计出的一类称作扫描探针显微镜的仪器（图5的照片就是用它拍摄的）不仅在观察方面颇具价值，操控分子世界也能力不凡。原子力显微镜（AFM）是这类显微镜中的一种，它允许研究人员探究分子的力学性质——例如分子的硬度或弹性如何。AFM能够抓起分子的一端，像拉橡皮筋一样拉动它。

设计马达

K.埃里克·德雷克斯勒是一位独立科学家，他领导着加利福尼亚州的前瞻学会，更是纳米技术领域最杰出的预言大师之一。德雷克斯勒的预见，大体上是设计分子尺度的自动组装机器人，能够一个原子一个原子地造出各种分子机器（也包括这些机器人自身），这种想法在公众对纳米技术前景（以及危险）的观感中颇具影响，在科学家之间反倒不那么流行。一些科学家担心，德雷克斯勒在原子组装机器人方面的想法忽视了原子结合时无可避免要释放出的热。而且，分子的形状虽各式各样，却并非随心所欲：我们无法保证，某种特定的纳米技术元件的分子尺度设计图就一定对应着一种稳定的原子排列方式，甚至都不一定是能够实现的排列方式。

德雷克斯勒于1986年在《造物引擎》一书中首次概述了他的思想，其中的主角（有时又是反派）就是纳米技术机器人。不过如今技术上已经实现从零开始造出可控的分子马达，虽然相当低级，但其本质已经是个造物的引擎。利用这种器件，我们就可以把分子

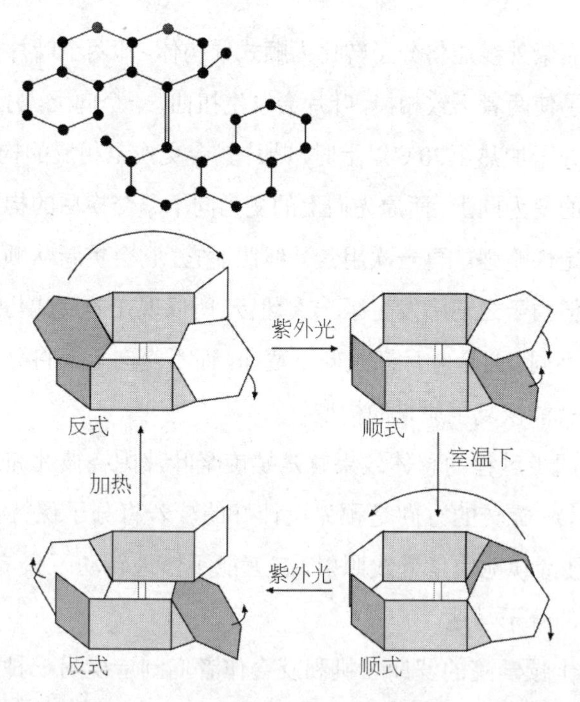

图30　人工制造的光驱动分子旋转马达。上部的图展示了它的碳原子骨架

尺度的直杆、横梁及其他建造部件移动至适当位置，再结合起来。

虽然马达蛋白使用ATP作为能源，一些研究者却想到，合成分子马达可以利用光作为能源。1999年，荷兰格罗宁根大学的本·费林加所领导的化学家课题组设计出一种旋转分子马达，其中转子能够受到光的驱动朝一个方向旋转。[①]他们利用的是感光异构的过程，即光诱导的分子两种不同形式（异构体）间的转化；两种形式的化学构成相同，但形状不同。

他们构造的分子含有两片相连的螺旋桨叶片状的部件（如图30）。起先，两片螺旋桨分别处于分子相反的两侧，称为**反式**异

① 费林加因该项研究成果获得2016年诺贝尔化学奖。——译注

构体。而紫外线能将分子转化为**顺式**异构体，即两个叶片处于同侧。为了使两者不致相撞，叶片会发生扭曲，一个向上，另一个向下。当分子加热至20℃以上时，叶片就会交换至相反的构型：原先向下的变为向上，而原先向上的变为向下。交换后的构型比原来的稳定性略强。再一次用紫外线照射它，它会重新从顺式变成反式。但由于之前刚发生了一次翻转，所以现在的反式与最初的也略为不同：两个叶片都是向下弯曲，而不是向上。将分子加热至60℃就能恢复至原来的构型。

这四步过程的总体效果就是螺旋桨叶片围绕彼此完成了一周的旋转，旋转的方向是预先设计好的。若将分子保持在60℃以上，且持续地用紫外线照射，它就能不停地转动下去，成为消耗光能的分子马达。

波士顿学院的罗斯·凯利及合作者们制造出另一种不同的旋转器件。他们构造出一种含三个叶片的螺旋桨，连接叶片的转轴上还有“制动器”可以阻碍螺旋桨旋转。若没有制动器，螺旋桨也能够转动，但会随机地向两个方向之一旋转。研究者们108 使用制动器，让叶片与制动器之间发生一系列化学反应，使它只向一个方向旋转。但截至目前，他们还没找到办法拉动螺旋桨旋转超过三分之一圈。

这两种器件都非常初级，无法完成有用的工作。但它们展示了，在原理上分子马达可能会如何构造。驱动凯利的分子马达需要一系列成键和断键，看起来很麻烦，可毕竟驱动蛋白和肌球蛋白的线性运动也需要相似的反应过程。在分子的尺度上，这类过程能够发生得很快，足以产生看起来流109 畅的运动。

天然纳米技术

合成分子马达要想与天然的马达蛋白比肩，还有很长的路要走。完全人工制造它们真的有意义吗？有没有可能调整现有的马达蛋白并与其他纳米技术相结合呢？一些研究人员已经将马达蛋白从细胞中分离出来，并用化学手段修饰它们，使它们能够完成新的任务。

1997年，普林斯顿大学的斯坦尼斯拉斯·莱布勒和同事们从马达驱动蛋白出发做出一种器件，能够将微管排列成有组织的模式。他们用化学手段将四个驱动蛋白连接起来，形成的组合体很像带四条腿的生物。将其与微管混合在一起，并加入ATP，这些驱动蛋白组合体就会一个接一个地拉动微管，最终形成星状结构（如图31），非常像细胞分裂第一阶段所形成的结构（参见第65页）。

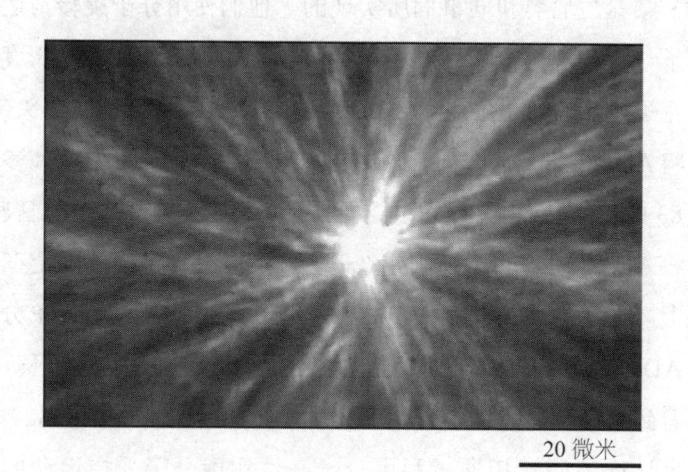

图31 对蛋白质进行修饰，制成半合成分子马达，可诱导微管组成星状结构 110

在西雅图的华盛顿大学，维奥拉·沃格尔和同事们利用驱动蛋白成功地在表面上朝选定方向推动微管。他们在表面上施以聚四氟乙烯涂层（PTFE）—— 一种不粘涂层，更通俗的名字是特氟龙——再将驱动蛋白分子附着于表面上。涂层是通过摩擦一块PTFE施加上的，于是在摩擦时，聚合物膜就沿摩擦方向形成了凹槽和凸脊的条纹结构。聚合物长链的方向应当与这些凸脊的方向平行。驱动蛋白分子就倾向于附着在这些凸脊上，也就是说它们会排成定向的列状。当微管传送的时候，这些列就成为线性的轨道：驱动蛋白分子像水桶传递接力那样一个接一个地传送微管。在细胞中，微管分子是“固定的”，驱动蛋白是运动的。而在这些实验中，马达蛋白固定在表面上，于是它们的行走运动就推动了微管。

迄今为止，生物分子马达与人工微工程领域最激动人心的融合发生在2000年底，是由纽约州伊萨卡市康奈尔大学的卡罗·蒙泰马尼奥和同事们所实现的。他们使用分子**旋转**马达推动了一个微观的金属螺旋桨叶片，叶片大约150纳米宽，长度约为宽度的十倍。我们之前曾提到过的ATP合成酶有一个头部，在将ADP转化为ATP时，头部绕着与膜相连的转轴旋转（参见第83页）。蒙泰马尼奥和同事们用金属镍蚀刻出微观的基座，并将ATP合成酶的头部固定在基座上，再将金属叶片固定在转轴上。在适当的条件下，ATP合成酶能够反向工作，将ATP分解为ADP，同时旋转。研究人员向转子提供ATP启动这一过程，继而看到它们在显微镜下开始旋转，平均每秒转五圈（如图32）。

这样的研究拓展了利用分子马达实现分子受控运动的前景，也为化学合成带来了分子尺度这一全新的层面。化学家不

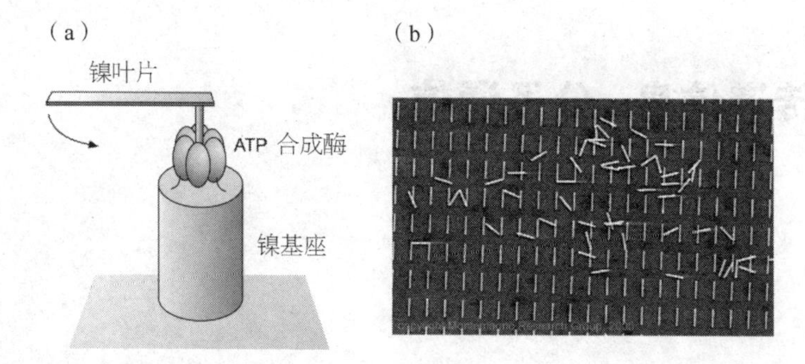

图32 (a)微观的金属叶片固定于旋转马达蛋白ATP合成酶上，在ATP作用下马达蛋白驱动叶片旋转。(b)这种组合体所组成的阵列。其中只有非竖直方向的叶片是按照预期正常工作的，这也说明并非每一例合成都成功

必再依赖于漂浮在溶液中的分子随机游荡、碰巧相遇，而可以精确地指定分子应当前往哪里。因为大自然已经设计出了用于这一目标的一系列奇妙的分子机器，所以我猜测分子纳米技术学家会越来越多地利用细胞中的小机器，而不是从头开始设计。这不仅是面向产生机械运动的问题，也面向能量产生、传感器、信息处理等很多领域。我们可能会见到生物学与从前截然不同的学科融合起来，比如机械工程、电子工程等等。正由于这个融合会带来学科独自发展所无法取得的成果，我们可以给它起个新名字，称之为**生物协同工程**。

111

112

第六章

传递信息：分子通信

我们每一个个体都是一片新世界。而从分子的视角看生命，一个个细胞非常像一座座城市，里面住满了分子居民。我们多细胞的躯体正是不同城镇之间协调合作的产物。细胞和细胞之间需要相互沟通，就像伦敦和利物浦、纽约和费城——信息会沿着电线传递，或是由信报员来送达。血液和淋巴循环系统像交通运输网，货物在路网上输送到四面八方。正如贝采里乌斯所言："这种生命的力量是身体机能与原料之间相互作用的结果。"

长久以来，分子生物学只满足于记录细胞的社会网络：推断哪种分子在对哪种分子说话，分子怎么来、怎么去。但这终究是不够的。我们还需要知道，分子说了些什么，信息是怎样一步步地传递的。药物化学家可以利用这些信息来研发新药。医学中一个基本的大问题就是搞清怎样参与体内分子的对话，从而拦截有害的或令人不适的信息，发出新的警告信息，阻止那些不希望发生的相互作用。

113 很大程度上由于生物化学领域的这些努力，化学本身的面貌得以焕然一新。化学家在生物学中看到的种种可能性点燃了他们的想象力。尽管化学工业有很大一部分致力于生产“消极”的产品——新型的塑料、水泥、粘胶、颜料、合成纤维——但药物分子却总是与它们有点不一样。药物的任务是要参与到一个动态的过程中，去影响细胞的活跃生命。它们像是在扮演剧中角色的演员——诚然，它们经常要靠扮演别人来发挥作用。如今，化学家也已经开始认识到，可以在纯合成化学系统中实现这样的动态活力。于是化学逐渐地较少关注单种分子的性质，而更多地去关注一群不同的分子如何作用——彼此结合或破裂，相互修改各自的趋向性，传递信号。化学开始成为一种**过程**的科学。

本书中我讨论的很多成果都支持这样的观点，比如研发分子太阳能电池、化学传感器、分子纳米技术、进行信息处理的分子器件等。这个领域内的很多研究都可以用**超分子化学**这一词来概括，意即超越了分子的化学——关于沟通着的分子的科学。

在本章中，我会首先探索生物学里分子负责通信的几种方式，之后再稍微领略一下如何对合成分子促进类似的群居共生。和以往一样，我们必须记住，尽管大自然是有启发性的，但它却也是吝啬和盲目的。生物所使用材料的广度很有限，而且生物倾向于无止境地为适应新的目的而给出还不错的解决方案，不会每次都完整地探索整条新的大道。正如喷气式飞机并不是放大版的鸽子，睿智的分子工程师从自然界中得到的是原理，而不是设计图。

114

分子邮件

当星罗棋布的小王国和小城邦为了合作共赢而一致尊重中

央当局的权威时，意大利和德国才成其为国家。同样，若不是细胞们类似行事，我们的躯体也就无法成为一个有一致性的整体。这意味着，体内必须要有某种机制，可以向全身发送关于动作的指示、命令和召唤。从大脑出发的神经信号就是躯体协调动作的一种方式。它们是躯体的电话系统。

而在**整个**身体范围内传送的常规信息则像是寄信，以分子形式作为群发邮件投入血液，这种分子就称为**激素**。激素的形态和功能多种多样。有的激素是大蛋白质，有的则是小有机分子。有的溶于水，有的则不溶（也就意味着要有信使分子在血液中携带它们）。有的传递紧急消息，比如“快跑！”有的则产生长期的效应，比如促进成长或者性征的发育。

所有激素都是**内分泌系统**的产物，这个系统包含一系列腺体，它们组成了对整个身体至关重要的调控体系（如图33）。我们已经看到了胰腺中产生的胰岛素和胰高血糖素如何控制血糖含量（参见第78页）。与此类似，细胞的代谢速率是由甲状腺所释放的甲状腺素和三碘甲腺原氨酸所调节的。这些激素能够一定程度地改变心跳速率，从而影响能量的产生和氧的消耗。

内分泌系统的控制中心是一个脑部的腺体——下丘脑。下丘脑与它下方的垂体相连，激素就从垂体出发被派遣到其他腺体。例如，代谢速率下降会刺激下丘脑，下丘脑向垂体发出一种叫作促甲状腺激素释放激素的分子。垂体继而就开始发出促甲状腺激素，触发甲状腺采取行动。

115

垂体释放的所有激素都是多肽，即像蛋白质的较小分子。例如，抗利尿激素通过调节肾脏中尿液的生成来控制体内的含水量。生长激素刺激细胞的增殖，在儿童期和青春期发挥重要

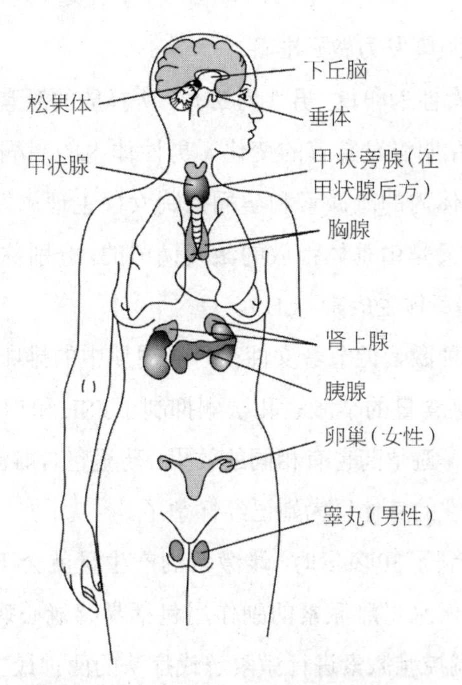

图33　身体的内分泌系统,即一系列产生激素的腺体

116

的作用。生长激素还能在需要修复时——如伤口愈合时,刺激组织的局部生长。

肾上腺制造一些重要的类固醇激素。这些不溶分子的碳骨架含有若干连在一起的小环。某些类固醇——如皮质醇——能够调节身体能量源的储存和利用,即葡萄糖到糖原的转化,及蛋白质切断成氨基酸。健美人士及运动员会利用这些激素(合法地或非法地)增加体重、锻炼肌肉。

正如你可能已经想到的,肾上腺素也产自肾上腺。在身体响应压力时,它和去甲肾上腺素一起,会迅速释放到血液中。这两种激素能使心率加快,血管扩张,提高对肌肉的供氧量,从而为肌肉最大限度发力做好准备。

性腺（女性为卵巢，男性为睾丸）所释放的激素能够分化性别，并在青春期引发发育的变化。男性体内的睾酮促进精子的产生。女性体内的雌激素和孕酮控制女性生理周期，而这两种激素的制造又是由垂体释放的激素调节的，分别称为卵泡刺激素（FSH）和黄体生成素（LH）。

上述两种激素调节着女性在月经周期中的排卵。在妊娠早期，血液中高含量的孕激素和孕酮抑制了FSH和LH的产生，从而抑制排卵。避孕药也有相同的效果：药里包含雌激素和孕酮，从而诱骗女性的身体以为她已经怀孕了。

当女性到了30多岁时，雌激素的产生量就会下降，特别是在更年期。低水平雌激素的副作用包括易感冠心病、骨质疏松等，这恰是调控雌激素进行激素替代疗法的两种最主要的疾病。这种治疗方法仍然存在争议，因为长期使用雌激素本身也会有副作用，包括易感乳腺癌及各种心脏病。

激发细胞

人体怎样解读激素信息呢？这依赖于信息本身的特性。有的激素可以穿过细胞膜，在细胞里与受体蛋白结合。这样就可以激发受体，刺激特定基因的转录，制造出细胞所需的蛋白质。这称作激素的基因直接作用机制，它作用于较小的不可溶激素，因而能够穿过脂类组成的细胞膜。

但很多激素最多只能敲敲细胞的大门而已，尤其是那些多肽和蛋白质分子的激素。它们会在细胞表面得到管家——受体蛋白的接待，受体蛋白的任务就是将信息传达至细胞内的其他分子。

和大多数分子通信类似，信息从激素传达到受体蛋白的方式颇为亲昵。分子之间毫不拘束，它们通过紧密的拥抱来传话。分子没有什么别的办法来相互识别，只能靠“触碰”辨识彼此，即受体与目标分子（底物）形状精确相符，嵌合在一起，如同钥匙配锁一般。细胞表面的每一个激素受体蛋白都有一个结合位点，形状塑造得恰好能与激素吻合。

尽管激素传递的信息多种多样，信号从细胞表面的受体蛋白传递到细胞内部的机制却几乎在所有情形下都是一样的。这个过程涉及一连串的分子相互作用，分子如接力般一个转化另一个。在细胞生物学中这称作**信号转导**。在信息进行接力传递的同时，相互作用还把信号放大了，于是单个激素分子与受体的对接就能在细胞里产生一个大动静。

这个过程是这样的：受体蛋白横跨整个细胞膜的厚度，激素结合位点在外表面伸出，而底部则从内表面露出（如图34）。当受体结合了目标激素时，一个形变就会传递到蛋白质下表面，使它成为一种酶。

这个酶所催化的过程是一种附着在膜内表面上的名叫G蛋白的分子“活化”。G蛋白的全称是鸟苷酸结合蛋白，即它能够抓住一个鸟苷二磷酸（GDP）分子。受体结合了激素后，就可以与一个抓住了GDP的G蛋白反应，G蛋白首先用鸟苷三磷酸（GTP，类似于富含能量的ATP）来替换GDP，接着整个分子一分为二。结合有GTP的那一半成为了一种酶，它会离开并激活另一种位于细胞膜内表面上的酶。这种酶通常是腺苷酸环化酶，它能将ATP转化为环状AMP（cAMP）。

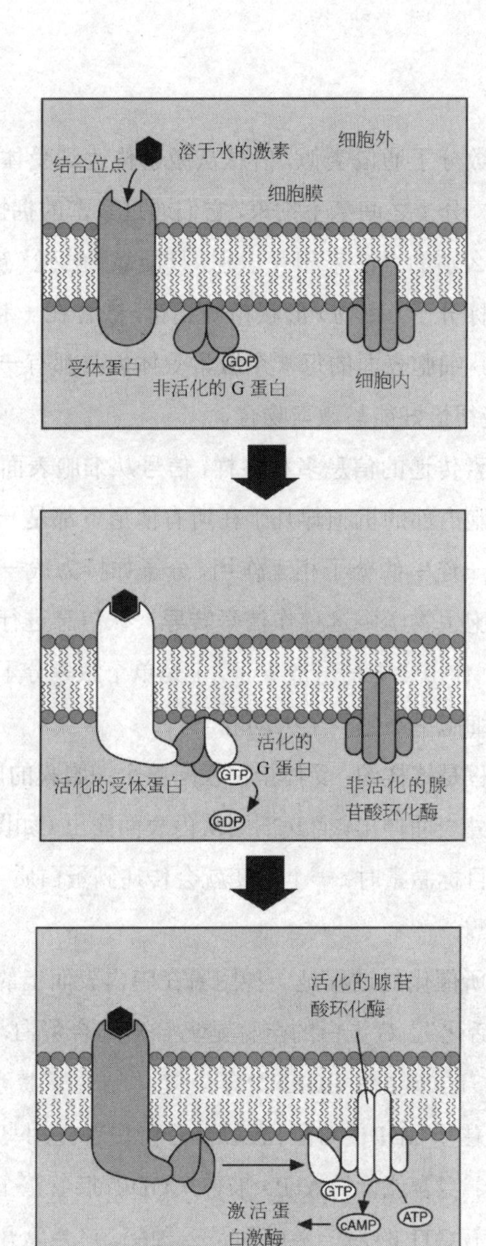

图34　G蛋白的工作方式

这个过程之前的所有参与者都卡在细胞膜上。但cAMP则能够自由地游动于细胞质中，能将信号传递到细胞内部。因为它从“第一信使”（激素）得到信号，并作为代理人传送到细胞社区中，所以它就称为“第二信使”。cAMP能够附着在称为蛋白激酶的蛋白质分子上，使得这种分子又变为活化的酶。大部分蛋白激酶能通过附加磷酸基团（一种称作磷酸化的反应）开启或关闭别的酶。蛋白激酶的行为启动了一连串的反应，每个蛋白激酶都可以作用于好几个酶分子，而下一个酶也可以作用于多个分子。这样一来，单个激素与受体对接就可以影响到细胞内部的很多分子，使信号得到放大。

这个过程可能听起来非常繁复，但它无非就是分子的接力。信号从激素传递到受体，再到G蛋白，经过一个酶后到达第二信使，继而到达蛋白激酶，依次继续下去。

信号转导的G蛋白机制是在1970年代由阿尔弗雷德·吉尔曼和马丁·罗德贝尔发现的，他们因此获得了1994年的诺贝尔医学奖。这种机制代表了信息通过细胞膜的最广泛的一种方式。有的激素可能对细胞进程起到的是减缓而非刺激作用，这种情况下活化G蛋白就可能针对目标酶表现抑制效应，而不是去激活它们。还有的时候，第二信使可能是非cAMP的其他小分子：比如，特定的活化G蛋白能够触发一种结合钙的蛋白质——钙调蛋白去释放钙离子。

利用G蛋白机制的也并非只有激素信号。我们的视觉和嗅觉同样涉及信号的转导，也会使用相同的转化过程。鼻腔的顶部排列着嗅觉传感器，称为嗅毛。嗅毛连接在神经细胞的末端，细胞能将信号传送到嗅球，即大脑的“嗅觉中心”。嗅毛的细胞膜上就分布着受体蛋白，它们专门结合进入鼻腔的特定的带气味分子。

嗅觉传感器有几百种不同种类，每种都带有一个结合位点，其形状恰好能接收某一种特定的常见气味。然而，我们能辨别的气味的范围要远远大于此，这是因为每种气味一般都是不同气味分子混合的结果。嗅球从不同的传感器获得混合的刺激，形成一种气味的“图像”，就像我们通过面部不同部分的总和来辨识一个人一样。

121

在气味信号产生时，G蛋白生成的cAMP会与一种称作钠通道的膜蛋白结合，打开通道使钠离子流进细胞。这就激发了一个神经冲动，会传导至嗅球。光刺激视神经细胞产生视觉信号也有着相同的过程。

我们的味觉大部分都可以归结于嗅觉系统。舌头上的味蕾只能辨别较低级的特征味道：甜、苦、咸和酸。而成熟的奶酪和新鲜出炉的面包带来的愉悦则大部分来自它们所释放的气味分子。

大脑的信号

虽然激素能够开启复杂的生物化学反应网，但它们所携带的信息却是相当粗糙的，只与生物成长和生存中最迫切的需求相关。而创造了西斯廷教堂、《魔笛》、相对论的分子间通信就是另一回事了。可归根结底，人脑也只是由分子组成的而已。

不过时至今日，大脑还是一个谜团，是科学中一大未解之谜。有的科学家认为大脑永远无法完全理解其自身，因为这个问题所具有的自我指涉的本性总会造成盲点。而有的科学家则相信，对意识的科学解释已经在地平线上显露出来了。不管怎样看待，大脑的秘密可能都远远超越了单纯分子的领地，而根植于与复杂的、紧密关联的信息网络的表现相关的问题之中。在这里我们看到了还原论的局限：尽管思想的分子过程我们已经描绘清楚了，但总体性的影响我们却一无所知。

大脑中包含很多脑细胞，或者称**神经元**，总数在10亿至1000亿之间。这并不值得夸耀一番，别的器官的细胞数量也与此不相 122 上下。但大脑与众不同之处在于这些细胞间通信网络的复杂性。每个神经元都有大约1000个对外的连接，于是整个大脑中就有多达100万亿的细胞间连接，差不多相当于1000个银河系里星星的数量。在这样的输送网络中，你瞬间就会迷失。计算机中集成电路的连接数量远不及此，因此也就难怪对于某些小孩子一下就能完成的任务，计算机即使算得再快也会悲惨地失败。

神经元会发送神经信号——其本质是电脉冲，信号沿着称作轴突的管状通道传送，传向另一个神经元。轴突的末尾是一系列枝杈，它们的尖端连接着其他神经元的细胞膜。这些连接处称为突触，在这里神经信号从一个神经元传递到另一个。神经元还会伸出许多较短的、繁茂的分支，称为树突，负责从其他神经元的轴突处收集信息。如果你喜欢，可以把轴突比作大脑的高速路，从一个神经元城市延伸到另一个。在突触的地方，高速路就进入了岔路，连接到树突，进入另一个城市的路网。

尽管轴突信号是电信号，但它们与电路中金属导线上的信号有所不同。轴突基本上是管状的细胞膜，沿纵向分布着允许钠离子和钾离子出入的通道。有的离子通道永远都是打开的；有的则有“大门”，根据电信号做出开关反应。还有的其实并不是通道，而实际上是个泵，积极地将钠离子排到外面，将钾离子拉到内部。这些钠-钾泵能够将离子向能量升高方向移动，即将离子从低浓度区域移向高浓度区域，它们能这样做的原因是从ATP得到了能量。

在“静息”状态，轴突内部和外部的钠离子和钾离子是不平衡的，造成膜两侧电荷的差异，亦即电位的差异：内侧液体比外侧多带一些负电荷（即“静息电位”）。当信号沿着轴突传递时，部分有大门的钠通道打开了，改变了离子的分布，使不平衡状态发生反转——内侧比外侧多带正电荷。此处的电位反转又进而打开了前方的钠通道，于是就沿着轴突移动下去。同时，后方的通道关闭，恢复了静息电位。电压脉冲或“动作电位”就这样沿着轴突传递下去（如图35）。

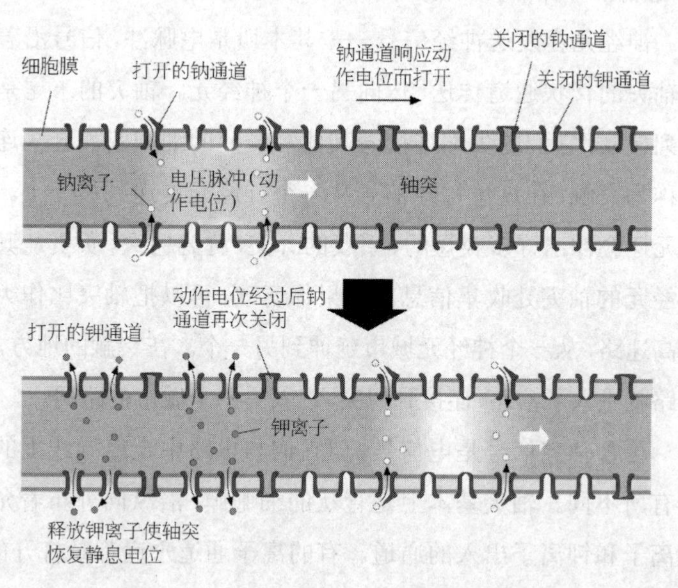

图35　电脉冲以打开和关闭离子通道的方式沿着轴突传送下去

在突触处，神经冲动会从一个轴突传到另一个神经元。信号一般先从电形式转化为化学形式。一种称为神经递质的小分子信使会将信号传过轴突末端膜与另一神经元膜之间的这段空间（称作突触间隙）。神经递质原本包裹在轴突细胞内小泡状的膜结构里，小泡与轴突细胞膜融合，这些分子信息释放到了突触间隙。神经递质移动到另一神经元细胞膜的外表面，与受体蛋白相结合。

124

有很多种不同的分子都可以作为神经递质。有的是简单的氨基酸，如甘氨酸和谷氨酸，或者是它们的衍生分子，如血清素和多巴胺。乙酰胆碱是在神经和肌肉连接处将信息从中枢神经系统传递到肌肉细胞的一种分子（参见第103页）。当乙酰胆碱与肌肉细胞上的受体结合时，受体就会转化为一个打开的钠通道。钠离子涌入细胞，改变细胞膜两侧的电压，打开电压控制的钙通道，这样就触发了细胞内部钙离子浓度的升高，刺激肌肉收缩。

乙酰胆碱体现了神经递质的普遍功能：打开或关闭离子通道，从而改变受体所在的膜两侧的电压。这就将化学信息重新转变为电信息。乙酰胆碱能直接完成这项任务，因为它的受体本身就是离子通道。其他的神经传递途径的作用方式与此不同：它们能使用第二信使将神经递质的信息转变为离子通道，和之前一样也用到G蛋白作为媒介。

G蛋白信号转导机制出现在如此多不同的场合，这不令人感到惊奇吗？其实不算。随着复杂的多细胞生物的进化，细胞也拥有越来越多专门的功能，而细胞的共同祖先则拥有较为普遍的功能。尽管必要的时候生物会适应环境，但对于特定的任务，久经考验的机制还是会保留下来。这也就是为什么我们会125 与酵母、细菌共享相同的一些基因。G蛋白通路就是一种将膜外化学信息传递到膜内的有效方式，还能在过程中放大信号。细胞的格言是：能用则用。

快感与痛感

神经信号转导是药物的一种常见作用目标——可能起到有益的作用，也可能有害，也可能令人愉悦，或者依赖于剂量而发挥所有这三种效果。神经系统是人体最脆弱的部分之一：如果神经冲动被阻断，我们就不能动了。很多动物能够产生毒素，攻击捕食者轴突上的钠-钾泵或者电压控制的离子通道，从而阻碍动作电位的传导，使捕食者麻痹。

与乙酰胆碱相像的药物分子会在神经肌肉连接处，和真正的乙酰胆碱竞争结合受体蛋白，从而影响肌肉动作。香烟中的活性成分尼古丁就是这样一种分子：它能够在肌肉上与特定的一类乙酰胆碱受体相结合，从而产生相关的刺激感受，使心率提高，瞳孔放大。但人们还没有完全理解，为什么这样的感受会令人愉悦。箭毒是一种致命的毒剂，存在于南美洲一种植物的树皮中，土著人曾经将它提取出来涂在箭头上。箭毒和尼古丁一样，会结合同一类乙酰胆碱受体，但并不激活它们，于是就阻止了肌肉的动作。中了箭毒的动物无法使肺脏扩张，会死于窒息。中世纪的一种毒药——毒芹也发挥相同的作用。

有的神经递质会刺激神经元，但也有的神经递质会起到使神经元平静下来的作用，即压制动作电位的火力。这些物质称为抑制剂，其中包括甘氨酸和 γ-氨基丁酸（GABA）。我们的意识是刺激和抑制之间复杂的交互作用，神经元会权衡它从各个邻居那里收到的不同信号，来决定自身是否应该激发。

LSD（麦角酰二乙胺）和墨司卡林这类致幻药物能够通过提高血清素的刺激效应来过度刺激大脑。毒药士的宁能阻碍抑制信号，引起肌肉失控地抽搐，导致十分痛苦的死亡。镇静剂则能够辅助抑制性神经递质的结合，或者（如酒精）干预刺激性神经递质的作用。

减轻疼痛的药物通常都与抑制性受体打交道。鸦片中的主要活性成分吗啡，能与脊髓中的鸦片类受体结合，从而抑制疼痛信号传递到大脑。大脑中同样也有鸦片类受体，所以吗啡和相关鸦片类药物既对大脑也对躯体产生作用。大脑中的这些受体本身是一类称作内啡肽的多肽分子的结合位点，内啡肽是大脑响应疼痛而产生的。有一些内啡肽本身就是功能极强的止痛药。

大麻素是大麻的活性成分，它同样能与大脑中的抑制性神经受体相结合，从而缓解疼痛。这些受体的天然靶物是称作内源性大麻素的分子，这种分子像内啡肽一样也是在响应疼痛信号时产生的。有一种和它紧密相关的分子称作油酰胺，这种分子似乎是自然睡眠的生化触发剂。

并非所有的止痛药都是通过阻碍疼痛信号而发挥作用的。有的甚至能直接阻止信号发出。疼痛信号由称作前列腺素的多肽引发，它是由受迫的细胞生成并释放的。阿司匹林（乙酰水杨酸）能与一种负责合成前列腺素的酶相嵌并加以抑制，于是从根源上切断了疼痛的呼喊。可惜的是，前列腺素还负责制造保护胃黏膜的黏液（参见第78页），所以阿司匹林的一项副作用就是

127 可能引起胃溃疡。

神经科学新近有一项惊人发现：非常小的无机分子同样能够作为神经递质。一氧化碳和一氧化氮都是双原子分子，它们就可以发挥这样的功能。在大剂量下，它们都是有毒的，因为它们会与氧气竞争结合血红蛋白。但“毒性离不开剂量”，少量的一氧化氮则能做一些重要的事情。它能触发血管的扩张，减轻心脏的压力。硝化甘油之所以能治疗心脏问题，就是因为它能分解并释放一氧化氮。伟哥这种药物用于治疗男性勃起障碍的基础也正是用一氧化氮来改善血液循环。

超分子化学

最近几十年，科学家对在合成系统中模仿细胞的部分分子通信过程产生了兴趣。他们的动机有很多。药物研发经常要编造一个良好的伪装，使得合成分子能够冒充成天然分子，且更容易与受体结合，从而阻碍或者引发一种生物化学信号。眼睛的视网膜细胞和嗅觉系统中都发生了信号转导，这启发了我们提出分子传感器的概念，用这样的传感器能以很高的灵敏度检测到光或其他分子。分子工程师正在研究嗅觉器官，期待从中获取灵感来设计“人工鼻子”，用以鉴别复杂的分子混合物。

“锁钥原理”是自然界高声歌颂的一项原理：分子相合则相聚。[①]若将这种“相识”转变为通信的过程，就需要让结合的事件能触发受体的某种转化，使信号能够继续传递下去。在生128 物当中，这种传递过程一般是催化，即结合使得受体转变为活化

① 这个比喻是德国化学家埃米尔·菲舍尔在1894年首次使用的，用以解释为何酶所催化的转化具有很强的选择性。——原注

酶。不过信号也可以通过其他方式传递：比如释放光或释放电子，或者（像乙酰胆碱受体一样）产生电化学势。

在超分子化学中，构建分子尺度上的人工信号转导过程是个常见的目标。这个领域正是受到生物学的启发而肇始的。1960年代，法国化学家让-玛丽·莱恩研究了所谓的冠醚分子，这种分子可以识别并结合特定的金属离子。莱恩的兴趣点是将钠离子和钾离子输送通过脂质膜。除了经由蛋白质通道和离子泵来传送外，还有另一种策略，即用分子把离子包裹起来，而分子“可溶解”于脂类细胞膜的内部。这种分子是天然存在的，名字称为离子载体。一个典型的例子是缬氨霉素，这是个环形的多肽，中间的空洞适合填入钾离子。冠醚则是模仿缬氨霉素的合成物质，它们同样也是环形，也能在中间的空洞结合金属离子。根据空洞与离子的相对大小不同，金属离子结合的紧密程度也有所不同。若空洞过大，则金属离子结合得较松散，会在里面“撞来撞去”；若空洞过小，金属离子就填不进去了。因此，我们可以调节冠醚使它适合特定的金属离子——换言之就是要展示分子识别的威力。

到了1970年代，莱恩和其他研究人员制出了各种形状和尺寸的合成受体分子，设计出的空洞能够适合范围很广的目标物，包括无机的，也包括有机的。这些“客人”分子被它们的主人用相互作用力抓住，这种相互作用比分子用来抓住自身原子的共价键要弱。这样客人就有可能被抓起来，然后再被释放。这正是缬氨霉素运送铁离子的方式：缬氨霉素先在膜的一侧捕捉一个离子，通过膜之后在另一侧把离子释放。超分子化学，本质上正是谈论用松散的联系把分子结合起来，又可以使它们解离成各个部分。

当冠醚抓起一个金属离子时，它们就会变形。冠醚单独存在时，是个非常松弛、软绵绵的环，像橡皮筋一样。当中心带有金属离子时，它就组织成较为稳固的结构，环上有了“之”字形的弯曲，也就是像个皇冠（如图36）。受体结合了标的物时，通常都会发生类似的变形。

如果我们的目标只是要结合而已，那么较大的变形就不太如意，因为受体的内部重组会让结合变得麻烦，较难实现。这也是为什么很多超分子的主人部分会设计得使它们在接收客人前就预先排列好，让结合引发的变形达到最小化。

但如果我们的意图是利用结合来触发信号的向下传递，那变形就常常是很关键的了。2000年，柏林洪堡大学的乌尔里克·薛尔特和同事们报告，发现一例主客结合所引发的巨大变形。他们制造了一种受体分子，你可以认为它是一系列“更小的分子”，包括两只胳膊，两条腿，还有一个像灵活的躯干的“传

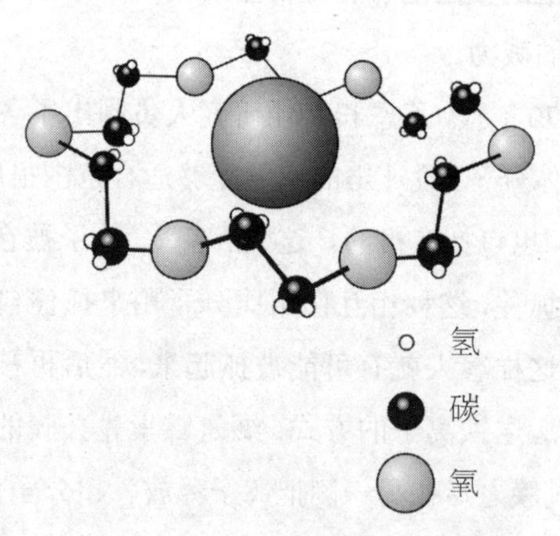

图36　冠醚是环状分子，中间的空洞可捕捉一个金属离子

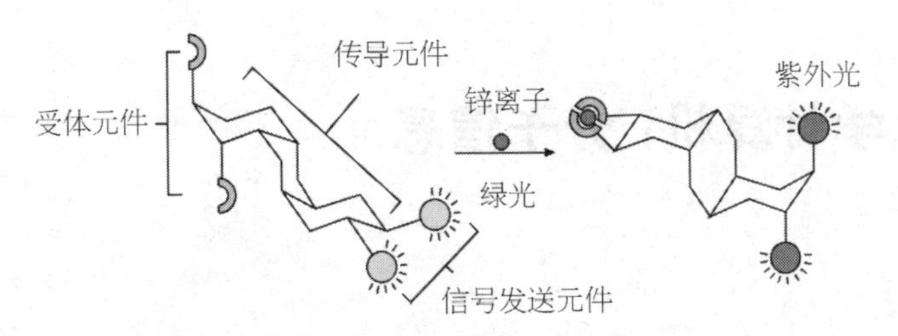

图37 一种人造的传感器分子，能够通过变形和荧光性质的改变，将与锌离子的结合转化为信号的发送

感器”元件把胳膊和腿都连接起来。若两只胳膊合起来围住一个锌离子，传感器元件就会翻个跟头，把两条腿拉开（如图37）。两条腿的末端有荧光基团，当它们距离增加时就会改变发射波长，从绿光变为紫外光。研究者们指出，这种受体分子显示出蛋白质受体在信号转导方面所具有的一些特性，与标的物结合时做出响应，一方面形状发生变化，另一方面行为也能发生变化。

人们通过分子工程，已经在好几种其他的合成受体上实现了由变形导致分子发光性质的改变。但要像G蛋白信号机制那样，通过识别与结合来转化分子的**催化**行为，则困难得多，因为这就要确保最终的形状恰恰就是催化剂能够工作的形状。而要将好几种分子组织成一条信号传递流，就更加困难了。无论如何，超分子化学家们的技能也在日益增长，如果我们很快就能看到人造分子通信系统协调地工作，像人体那样精妙和谐地调节着自身的领地，那也不值得大惊小怪。

第七章

化学计算机：分子信息

在最后，我们都留下了相同的疑问：生命是什么？至少在这本书里，我不会回答它。但是薛定谔给出的答案——负熵（参见第74页），尽管有一些缺陷，却不失为把握住了一丝的真理。负熵是生命所必要的特征，但不是充分的，它在混乱上推行着秩序。混乱就是死亡。如果细胞不能明晰地收发信息，如果细胞不能在正确的时间完成任务，如果细胞膜失去了组织性，如果蛋白质不能折叠，那么生命将无以为继。我们正是荒蛮世界中的秩序绿洲。

秩序从何而来？无机的物质也能够自发组织：想想成排的马尾云，或者是风吹沙土形成的规则纹路。似乎给系统提供能量防止达到稳态平衡，这类“自组织”就很有可能自发出现，而它在生命的秩序性中也发挥着作用。但仅仅如此还不够。细胞能复制染色体并一分为二，能反复制造出功能性的蛋白质分子，还能从单细胞的受精卵发育成多细胞的莫扎特，这期间所需要的协作配合不能只靠描画天空和沙漠的这种“盲目”的绘制过程。必须有一只手更为牢固地握住方向盘才可以。

我们都听过这只引导方向的大手的名字。它就是DNA，一串分子珠链，切开并打包成46捆小小的X形的染色体。人类的基因组——我们的全部DNA——常称作“生命之书”。在我写这本书时，科学家们刚刚结束了这部大书第一份草图的解码工作：他们绘制出了每条染色体上分子信息的大体细节。

关于人类基因组计划，社会上流传着一些不太谨慎的说法。比如有传言说，一个技艺足够高超的工程师仅用其中的信息就可以造出一个人。这是无稽之谈。人体中充满了各种并非靠基因组编码的分子，基因组所编码的只有蛋白质而已，甚至还编得有点杂乱、不完整。关于组成细胞膜的脂类，基因组什么都没说，更没提到这些脂类会如何在物理动力的作用下聚集成层状、环状和球状。基因也不会告诉我们神经信号如何工作，大脑如何利用精确定时的电脉冲序列来编码思想和感觉。也没有哪个基因是关于骨骼、牙釉质的。说基因组是细胞之书，就像是说字典是话剧《等待戈多》之书。它全部都在里面了，但你却无法由此推导及彼。

话说回来，基因组**的确是**一部分子形式的指导手册。它告诉我们如何制造蛋白质，而蛋白质编导着生命壮阔的分子大戏。在这一章里，我想多谈一些这个剧本的本性——它是如何阅读，又是如何演出的。而我最终的目的还要更加宽泛。对于分子科学家而言，遗传学谈的其实是分子非常奇妙而深刻的一个特点——能够携带并传递信息。普林斯顿大学的理论生物学家约翰·霍普菲尔德指出，这是生物学用“存在性定理”来启发化学家的众多例子中的一个。他说：“数学家使用‘存在性定理’这个词，指的是证明他们想要构造的某种函数确实存在，而不是不

132

133 可能的。从这种意义上讲，观察到鸟在飞就为工程师提供了一个存在性定理，证明我们能够设计出会飞的机器。”

根据同样的道理，遗传学向我们展示了利用分子进行计算是可能的。计算无非就是信息的存储、传输和处理，而基因机器可以完成所有这些。让-玛丽·莱恩说：“存在着一种‘生命有机体的分子逻辑’。”

这其实是上一章的推论，我们在上一章中已经看到，分子之间可以相互通信。而真正的分子逻辑含义更加确切：它不仅要求一个分子可以影响另一个分子的行为，还要求它们能以清楚严格的方式传输并操作经过编码的信息。计算机就是这样工作的：让数据在半导体和磁性材料制造的开关和存储器间进行传输。

用分子来计算仅仅是信息走进分子科学的其中一面。从更宽泛的方面来讲，化学家已经熟知了给分子编程让它们表现特定行为的思路；可以把性质编织到分子结构体中，就像对机器人编程、写入一系列指令那样。莱恩说：“其前景……是一种关于信息化物质的更广泛的科学。”这样的化学是真真正正的一种崭新的科学，在许多方面都与传统的制造实用物质的化学截然不同。这种科学是关于更加积极的“演化”的，而不再仅关乎消极的“存在”。它正在发生着，但我们还不知道它会带我们走向何方。

细胞从何而来

每一部书都用一种特定的语言写成，基因组也不例外。基134 因的语言是种简单的编码，它包含的字符是四种核苷酸分子，这些核苷酸分子就是DNA分子珠链上的珍珠（参见第42页）。每个核苷酸分子包含一个所谓的碱基，信息就编码在碱基当中。

DNA中有四种碱基：腺嘌呤、胞嘧啶、鸟嘌呤和胸腺嘧啶（分别记作A、C、G、T）。DNA是核苷酸单元组成的线状聚合物，所以编码的信息就可以表示成四种字母组成的线状字符串。字符串可能会包含如下的一段：

GTGGATTGACATGATAGAAGCACTCTACTATATTC

只包含四种字母的字母表看似非常局限，不适于书写复杂的信息。但如果我们将这个序列看作一种密码，而不是严格地看作一种字母表，那么你想要多复杂它都能做得到。比如，我们可以将每个罗马字母表示成若干碱基的序列：GTG表示“a”，GAT表示“b”，等等等等。长度为三的四字符序列一共有64种，多于整个字母表的字母数量。使用这样的密码，我们就可以用AGCT的字符序列来书写《圣经》。

《圣经》里的信息对细胞来说没什么用，细胞需要的是能够用来制造蛋白质的信息。蛋白质长链如何折叠是由它的氨基酸序列所决定的（参见第41页），因此氨基酸序列就唯一地规定了制造蛋白质所需的“信息”。DNA编制这种信息所用的密码正如我们前面所提示的：三个碱基一组代表一种氨基酸。这就是遗传密码。[①]

① 需要指出的是，关于遗传学的描述需要作许多简化，这里就是一例。我们已经讲到，有些蛋白质非常重要的组成部分并不含氨基酸，而是由其他化学基团组成的，比如血红蛋白中的血红素单元。这些基团称作辅基，它们是其他的酶制造出来的，在蛋白质长链造好之后再加入其中。而没有辅基的单纯的蛋白质链称作脱辅基蛋白。一般而言，在加上修饰之前这些蛋白全无用处。要真正地推断蛋白质的结构和形状，除了知道氨基酸序列，即编码它的基因序列之外，我们还需要知道之后在脱辅基蛋白上进行操作的那些蛋白质的特性和功能。——原注

人们至今尚未完全理解一个特定的蛋白质序列会如何折叠它的链。也就是说，我们还不能够仅凭基因的序列就推断基因的功能（虽然我们有时可以大致猜到）。人类基因组的第一幅草图里面还充满着目的不明的基因。

不过细胞中信息流动的原理我们完全理清了。DNA是关于蛋白质信息的手册。我们可以认为每个染色体都是独立的一章，每个基因则是这一章中的一个单词（它们可是非常长的单词！），基因中的每个碱基三元组是单词中的一个字母。而蛋白质就是单词翻译出的另一种语言，新语言的每个字母是一个氨基酸。一般而言，只有当基因语言翻译出来以后我们才能理解它的含义。

DNA是一种双链的聚合物：两条链彼此扭曲盘旋，形成双螺旋。每条链都是一个核苷酸长串，信息就编码在里面。但两条链并非全同。这条链上的碱基可以和那条链上的碱基之间形成氢键（参见第41页），两条链就像拉链一样通过氢键相互嵌合。虽然所有的碱基都能形成氢键，但它们有特定的结合选择，A和T相结合，G和C相结合。所以DNA双螺旋包含的是互补的序列：每当A出现在这条链上，T就出现在那条链上，依此类推。这就意味着每个基因都写成了两个版本，以镜像的语言彼此呼应。

碱基这种两两成对的特性是它们的形状所决定的。碱基A和碱基G是相似的分子，C和T也是相似的分子。于是A-T组合体与C-G组合体的形状和大小大致相同。碱基对在两条螺旋链的内侧连接，像螺旋楼梯的台阶。只有台阶的尺寸都一致，两条链才能平顺地盘旋下去。若A与G结合就会鼓出一块，发生扭曲变形，破坏两条链的结合。同样，若C与T结合就会陷下去一 136 块。另外，台阶中氢键的位置决定了A–C和G–T的组合也是不成立的。因此，其实是搭档间吻合的**互补性**造就了碱基两两成对的偏好。

生物信息流动的一个关键要点是：数据的传输通过分子识别过程进行，确保信息的每一部分都得到正确的解读。

当细胞分裂时，DNA会进行复制，也就是基因组得到复制。因为两条链完全互补，所以它们都可以作为模板来组装新链。如果A总是优先与T配对，且依此类推，一条“赤裸的”单链就能引导游离的单个核苷酸按正确的顺序连成一线，形成一条互补链。

为了扮好模板的角色，双链首先会在特殊的酶的作用下拆成两条单链。然后沿着暴露的单链，互补链就被组装起来：称作DNA聚合酶的酶就催化了新核苷酸的加入。于是两组新的双螺旋都各含原先双螺旋中的一条链。

尽管酶能够帮助这一过程进行，但复制过程所必要的信息都已写入DNA模板当中了。1980年代初，加利福尼亚州索尔克研究所的莱斯利·奥格尔和同事们展示了在没有酶辅助的条件下，单体核苷酸也能够基于互补核苷酸的模板组装成聚合物。例如，一段八个C组成的RNA核苷酸序列，可以作为模板组装起八个G的核苷酸序列。不过奥格尔也不得不在其中做一点手脚，用的G核苷酸是通过加入活性化学基团“激化”过的，于是帮助它们连接起来。 137

这种模板辅助的聚合本身并不是复制：新链与模板是互补的，而不是全同的。第一例真正的人工分子复制是在1986年由

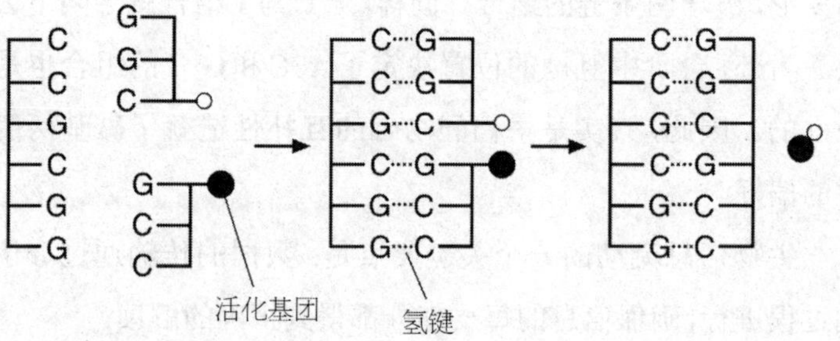

图38　长度为6的核酸中的分子复制

德国化学家君特·冯·凯德罗夫斯基报告的。他使用同样的模板组装过程，但选择的是自补的模板，即自己与自己形成互补。他的模板是个含六核苷酸的DNA分子，序列为CCGCGG。因为双螺旋两条链的头尾方向关系是头对尾、尾对头，两者逆向对接，所以模板的互补序列与自身完全相同。君特·冯·凯德罗夫斯基从两种三核苷酸的片段出发，组装成模板的互补链，其中同样需要活化帮助它们连接起来（如图38）。

谬误和冗余

在出版这本书的某个阶段，我会从出版社收到校样——最终成书页面的初级版本，由我提供的原稿编辑而来。（但愿）它将会是我所写内容多多少少较为忠实的转录。但毫无疑问，其中总会散布着零星的小错，可能是打字错误，也可能是文件读取故障导致的。作者们对此习以为常，因为要复制一份很长且很复杂的信息总难免引入一些错误。

基因的转录（DNA复制为RNA）和翻译（RNA复制为蛋白质序列）过程同样如此。分子也不能永远都完美地识别，偶尔会有一个错误的核苷酸或氨基酸插入链中。大概每20个蛋白质中就有一个的制造过程出现差错。

这要紧吗？总的来看并不要紧。我和出版社不太可能在这本书付印之前找出所有打字错误。但很可能里面的错误都不太严重，不至于让你无法理解我的意思。类似地，在蛋白质中，链的大部分都是充当脚手架的作用，只是为了将个别要执行催化任务的氨基酸残基放在正确的位置上。所以发生在脚手架上的各处错误可能都不严重。有时一个错误也可能会导致产生的分子完全失效，但细胞对于任意特定任务会制造不止一种酶分子，而往往会造出几十种甚至上百种酶分子来执行这个任务，所以即使有一两种废品也没关系。

我这里所讲的是**随机性**错误。而**系统性**错误就要严重得多，它产生在生物信息流的上游位置，更靠近于信息存储的根源位置。若RNA分子转录有错，就会产生出上百个错误的蛋白质。因此，会有一些酶专门仔细检查转录过程中有没有产生复制差错，把错误的频率降低至大约万分之一。

但即使是转录中的错误也很少会造成很严重的后果，毕竟RNA分子很短命，细胞也总能造出更多的RNA来。而DNA里出现错误就糟了，因为一旦错误产生就没办法再去纠正。若基因中一个核苷酸放错位置，这段基因产生的RNA以及这些RNA再产生的蛋白质就都会含有相似的错误。更糟的是，由这个基因有错的细胞中分裂出去的后代所有细胞都继承了相同的缺陷。如果配子——精子或卵子细胞——带有基因缺陷，那么缺陷将传播至该配子所繁衍出的所有后代上。这就是为什么DNA复制时需要“校对”酶来极端认真地审核，它会保证平均每10亿

139

个碱基中混入错误的数量不大于一。若缺少了这些校对分子，每产生一个新细胞就会得到大约1000个有缺陷的基因。

能遗传下去的错误，即制造配子时DNA复制发生的错误，就称作突变。一旦突变产生，它们就会沿着系谱树从亲代一直传到后代。突变是一些基因相关疾病的原因，如囊性纤维化疾病等；突变还会导致一些基因相关的易感体质，如易感癌症和心脏病等。尽管突变会导致这些可怕的后果，但它同时也是生命中的调味料。实际上，正是有了基因突变，才会有了我们人类的存在。如果在早期地球上热汤之中的低级单细胞生命体从不偶然发生突变，总能不带任何错误地复制相同的DNA，那么就不会有进化，也就不会出现更复杂的生命。

当出版社给我寄来校样时，文本必然会发生一点变化。校样中常出现一些我当初并没有写过的词语。但这并不是错误，而是完全合理的。它们其实是编辑所作的改动，而且我能肯定新的文本比我的原文更易于阅读和理解。

在1970年代中期，人们惊讶地发现基因同样也需要编辑。从DNA模板上直接脱离下来的RNA转录副本并不适于翻译成蛋白质，它含有很多无用的信息。这些“初级RNA转录副本”更像是语句中被随机插入了其他的语句碎片。RNA分子需要进行大量的编辑，才能表达清楚的信息，适于翻译。

这些插入的无用信息称作内含子，有时它们甚至会占据基因的大部分空间。它们并不用来编码蛋白质，所以也称为非编码序列。酶会在RNA初级转录副本中剪掉内含子，然后将编码区（称为外显子）的两段拼接起来。

“细胞之书”中遍布着杂乱的内容和无聊的重复，而上述只

140

是其中一种形式。人们认为，整个人类基因组中只有百分之二到三的部分是用来编码蛋白质的。有的序列发生重复是有理由的。每个人的染色体都以TTAGGG重复约2500次结尾。这些片段称为端粒，人们认为它是用来保持染色体稳定的。细胞每分裂一次，它们就会被截短一次，这种侵蚀在老化过程中发挥了作用。但也有很多其他的重复序列并不具备有用的功能。转位子是一种能在基因组上跳来跳去的重复序列，每离开一处时就会留下一些副本。人们认为，这是一种居住在我们体内最核心处的基因寄生物，它们唯一的目的就是复制自己。内含子可能就是丧失了移动能力的远古转位子残余。

剪切、拼接、复制以及合成核酸的蛋白质机制为我们提供了基因生物技术的重要工具，让我们能够操纵基因组。比如限制酶就是能够识别特定DNA短序列并在该处切开链的蛋白质。连接酶能把DNA的松散端头连接起来。利用DNA聚合酶，我们可以在试管里无限地复制DNA片段。通过加热我们可以解开双链，进而用于模板复制。反复循环进行复制和加热，就可以成指数倍地增加DNA。这个过程就称为聚合酶链式反应（PCR），使用的酶是从在温泉中生活的细菌体内提取的DNA聚合酶。这种细菌的酶已经在进化中获得耐高温的能力，因此其DNA聚合酶不会被反复的加热破坏。

这些工具使得科学家们能够“改写细胞之书”，意即向生命体的基因组中插入新的基因。农作物科学家希望把抗虫、抗涝、抗除草剂基因植入植物中，还希望结合改进作物风味、提高增长率的基因。但这也有潜在的风险，具体而言，比如抗除草剂的基因有可能会从农作物体内转移到杂草的体内，产生“超级杂草”

的新品种。人们尚不知晓这种基因跨物种“横向”转移的概率有多大。

有些人反对基因工程，认为篡改基本的生命材料DNA是违背伦理的，无论对象是细菌、人类、西红柿还是绵羊。这种反对意见是可以理解的，而且若以它不科学为由来驳斥就太傲慢了。不过这种意见的确与我们对生命分子基础的认识不太契合。一旦意识到我们的基因组成是多么随机——即便称不上任性，恐怕我们就很难再继续把它看作神圣不可侵犯的。我们的基因组里到处都是寄生着的废品，充斥着30亿年进化的残余。这套不成体统的文库里面似乎并没有什么优雅的、值得尊崇的东西，而真正值得尊崇的应该是蛋白质，一群群勤勤恳恳的蛋白质从长篇的废话中努力地筛选出有意义的片段。这一整套工作完成得如此之好，的确令人惊讶。但正如大多数生命现象一样，这也只是得过且过的办法，效率和整洁并不是重要的问题。

建筑设计图

DNA是“信息物质”的一个极端例子。它以非常特殊的方式遵循程序组装起来，每个核苷酸都与它的互补位置相匹配，一直能让上千个碱基对相互匹配。这种程序化自组装正体现了超分子化学的一个重要目标。原子自身相互之间的区别并不太大，但超越原子这种基本的建造单元而考虑制造分子，超分子化学家就可以给他们的砖块和马达编制程序，加入更多的引导信息。

DNA提供给我们的并不仅仅是程序化自组装的存在性定理，它也提供了编制程序的具体方法。对于比细胞的双螺旋远为复杂的结构，我们完全可以利用碱基互补原理把DNA支架拼接起来。

组约大学的纳德里安·希曼就探索了这样的概念。他使用生物技术中的酶工具将DNA剪切并拼接成漂亮的形体，比如他造出了一个笼子状的多面体——立方体加个截角八面体（如图39）。形体中的棱都是DNA双螺旋，每个顶角处都有三条螺旋会聚。这些会聚处都是精心编织而成的：双螺旋的两条链在此处分离，走向不同的棱，与新的链互补组成双螺旋。对于会聚处的情形，希曼和同事们先精心设计好DNA序列，合成序列然后再拼接。

巧妙之处在于该怎样把三棱会聚点组装成三维几何结构的分子物体。希曼让棱都带有“黏性端头”，在端头双螺旋的其中一条链长于另一条。于是暴露在外的未配对碱基在遇到互补链时就能够去配对。这样一来，端头就具有了带选择性的黏性，可

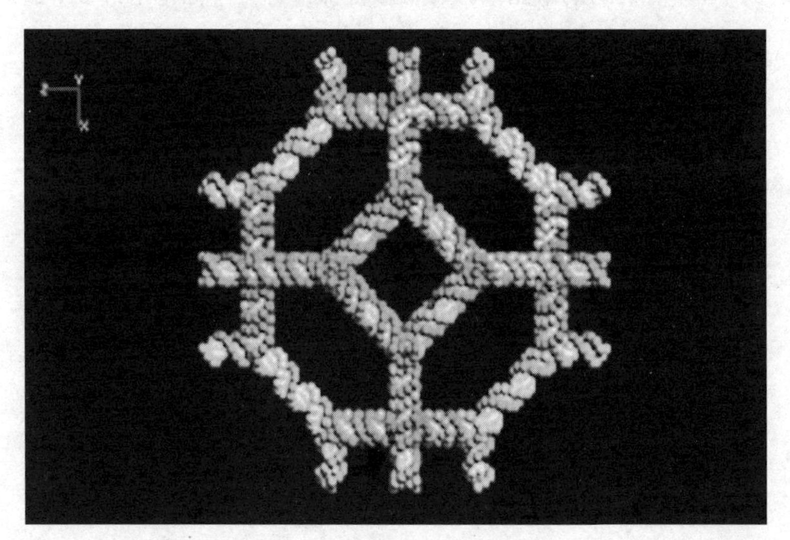

图39 用双链DNA组装成的多面体状分子

以与别的端头结合，于是整个结构就可以程序化地从零部件开始建造起自身。一旦黏性端头通过碱基间的氢键与别的链结合起来，接合酶就会在它们之间锻压出强键，保证骨架的稳固。

目前为止，这些分子建造其实还只是愚笨的技巧表演，只不过展示了对分子识别实现惊人的控制，从而建造起纳米尺度的结构。但希曼提出，在他这种实验中，DNA框架有可能充当脚手架，有效地组装起其他分子和材料。例如，我们有可能在DNA链外部覆盖上银，于是就把它们转化为导电的分子导线。那么，是不是有可能某一天通过对导线进行“基因”编程，建造出微型的、按特定模式连接起来的电路呢？

此外，带黏性端头的DNA还能够有选择性地粘接成分子尺度的物体。伊利诺伊州西北大学的查德·米尔金和同事们利用这个思路将金的小微粒组装成团簇。小颗粒原本只有几个纳米大小，即所谓的纳米晶体。每个颗粒都带有一个标记单链DNA，但标记序列没有互补关系，所以微粒保持分离状态。研究者们向其中加入一种单链DNA，它的两端分别与标记序列互补，于是纳米微粒就被连接到了一起。结果得到的团簇能够强烈地散射蓝色光，于是溶液就变成了葡萄酒的颜色。米尔金和同事们现在正在研发该技术的商业用途，希望成为特定DNA单链序列的简单的可视化检测方法，而这种检测方法正是基因分析中普遍的需求。

有的研究人员希望通过将金属或半导体的纳米晶体组装成有秩序的阵列，来制造更小的电子器件，远小于当今用来制造硅芯片的传统微细加工技术所制造出的器件。半导体纳米晶体可以用作存放电子信息的存储元件，它们还能与光相互作用，用来制造光学信息处理器件。利用DNA连接来程序化地组装纳米晶体，有可能成为排列电路模式的一种方法。得克萨斯大学的安杰拉·贝尔彻和同事们的研究还提示了另一种可能的途径：他们利用的是蛋白质的分子识别性质，而没有利用DNA。他们研发出一种很小的多肽分子，能够识别不同种类的半导体，并粘接在其表面。这种多肽可以“感觉到”半导体晶体表面的原子是怎样组织的。也许有一天，可以给基因工程马达蛋白（参见第五章）装上这种识别特定半导体的多肽手臂，于是就能将纳米晶体拖拽到分子建造的位点附近，并把它们排列成电路的模式。

分子逻辑

自从计算机在1940年代发明出来，新机器的计算能力大约每18个月就翻一倍。这种趋势就是以1965年首次提出者、英特尔的共同创始人戈登·摩尔的名字所命名的摩尔定律，它是由电路微型化驱动的。随着人们能够将更多的电路元件压缩在既定的空间中，计算能力就得到了提升。但如果摩尔定律还要继续维持25年以上，电子器件就必须减小到纳米量级，即分子的尺度。

但还没有人知道应该怎样去实现它，因为在这样的尺度下，集成电路的劳力——硅晶体管就会发生泄漏，不能用作开关。为了继续让计算机变得更快、更强，有一派思想正在不断壮大，认为元件必须变成单个分子。这种前景和传统的信息技术大不相同，投入在其中的人们堪称胆大、狂热的投机者。

但这也并不是个新的想法。早在1974年，美国化学家马克·拉特纳和阿里·艾维瑞姆就提出了一种单分子整流器（只允许一个方向电流通过的器件）的设计。仅仅数年之后，学者们就发现了能导电的碳基聚合物，并开始寄希望于这种材料的单分子能够用作“分子计算机”的导线。**分子电子学**这一领域就此诞生。

但在接下来的约十年中，再没有更多的进展了。它只是一种超前于时代的思想，合成、排列、检测这种梦寐以求的分子器件的实验手段却是缺失的。而近些年来，好几条路径汇聚到一起，使这一领域复兴，并且分子电子学和它所附带的分子计算也最终得到来自重要人士——制造计算机的公司——的认真关注。

分子信息处理技术的一个核心要素就是开关——能替代晶体管的器件。用最初等的语言讲，开关能够处于两种不同的稳定状态——“开”和“关”。晶体管在“开”的时候导通电流，“关”的时候阻断电流。不过，开关之间只有在相互连接起来时才能有信息处理的用处，能够彼此通话，前后传递信息。而用分子实现这个目标很难。

1999年，加州大学洛杉矶分校的詹姆斯·希斯及合作者们宣布了这方面的成果，这是一项与计算机巨头惠普公司的合作。他们基于有机分子实现了数个开关的互连，制造出了电控制的逻辑门。

在计算机电路中，信息是以二进制形式编码的，即1和0组成的序列。信号1对应于一定电压的电脉冲，信号0对应于零电压。整个电路中只发送这两种信号，没有信号1/2或者信号2。数据编码成1和0的序列，正如同DNA将信息编码成核苷酸碱基序列。二进制编码比基因编码还要简单一些：它只有两种字符而已。编码信号中的每个信息单元——1或0——就称作一个二进制位，或者比特。

计算机使用逻辑门来操作二进制信息并进行计算，所谓逻辑门就是进行决策的器件或电路。逻辑门接收一个或多个输入信号，并发出一个或多个输出信号。而输出的信号依赖于输入的内容。以“与”门为例，它接收两个输入比特，产生一个输出比特。若两个输入均为1，则输出也为1；若其中任何一个输入为0，则输出为0。对类似这样的逻辑门进行简单地组合，就可以执行算术：例如，以二进制的形式读入两个数，产生的输出编码了两者之和或两者之差。

希斯和同事们用一种称作轮烷的分子开关构造了“与”门。轮烷是一种两分子的组合体，由一个环穿在一根棒上。棒的两端固定有大螺帽部件，防止环掉出。棒被设计得能够吸引环，于是两种分子混合时就自发套在一起。这之后再加入端头的螺帽。希斯在加州大学洛杉矶分校的同事弗雷泽·斯托达特[①]，早在1980年代于英格兰的谢菲尔德大学就研发出一种技术，用以制造这种分子组合体。

研究者们将轮烷在金属电极上排布成一层，并在其上沉积很细的金属导线。在导线上施加电压，分子就从低导通状态切换到高导通状态。依附于单个导线上的数千个分子就组成了一个单个的、可切换状态的器件。研究者们将几个这种装置连接起来，制成了一个“与”门。

他们说，从原理上讲，应当可以用单个可切换状态的分子来制造这种器件。但在单个分子间进行电路连接很困难，也很难测量其间的微弱电流。但这也并非完全不可能。美国的马克·里德、詹姆斯·图尔和同事们就测量了连接两个金电极的单条“分子导线”上的电导率。

① 斯托达特因轮烷方面的研究成果获得2016年诺贝尔化学奖。——译注

弗雷泽·斯托达特与博洛尼亚的温琴佐·巴尔扎尼以及其他同事们合作，展示了单分子制成的另一种不同的逻辑操作——“异或”门。和“与”门一样，“异或”门也有两个输入和一个输出。当输入信号不同时（0和1，或1和0），输出信号为1；当输入信号相同时（0和0，或1和1），输出信号为0。研究者们观察到准轮烷也有这样的行为特点。所谓准轮烷，也是一种环套棒型的分子组合体，但没有棒两端的阻挡，所以环有可能从棒上滑落。

这种装置的输入不是电信号，而是化学信号，产生的是光信号输出。也就是说，它能够根据“化学信号”究竟是“开”是“关”，即两种化学物质是否存在，来改变自身的光发射行为（荧光）（如图40）。这与细胞表面受体蛋白的工作方式颇为相似（参见第119页），受体蛋白能够根据它们是否结合了目标分子来发出某种信号。

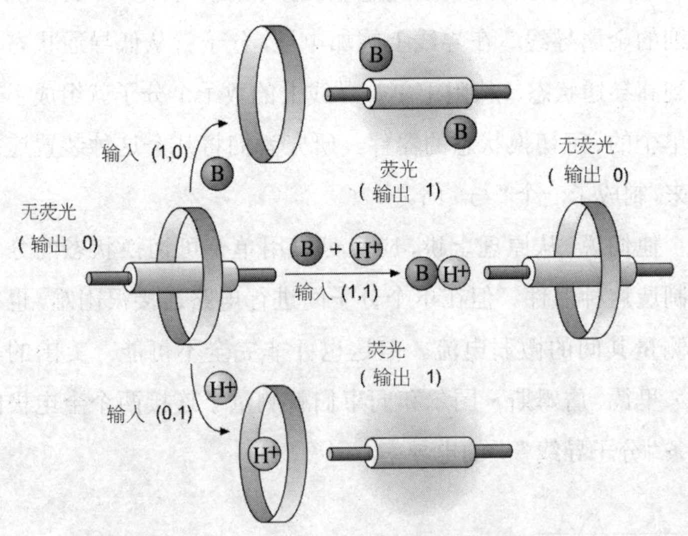

图40　名为轮烷的环套棒型分子组合体执行分子逻辑

利用相似的原理，北爱尔兰贝尔法斯特大学的A.普拉桑纳·德·席尔瓦和内森·麦克莱纳根结合了两种分子逻辑门，使它们能执行基本的算术。也就是说，他们能够使用分子来计数并完成简单的求和，诸如1+1=2。

当然，从计算一加一到制造可与硅基器件计算机相媲美的新计算机，还有相当长的距离。但这些研究已经展示了一个重 148 要的原则：分子确能用于计算，且是在单分子器件的层面。它们最终让我们看到，分子计算机并不仅仅是个狡猾的广告。

我们越近距离地审视关于新计算机的想法，就越能够看到它与人体所面临的难题的相似之处：怎样随心所欲地排列分子，怎样传输并放大信号，怎样使导线在两个开关装置间生长出来（像神经元那样），怎样处理差错，怎样控制事务之间的相对时序。或许未来的计算机工程师也需要学习很多的生物学知识。

DNA计算

似乎是为了把DNA发挥到极致，近几年来一些科学家展示了DNA也能够执行计算。这让我们转了一大圈之后又回到起点，因为我在一开始就说到，DNA提供了分子计算的一种存在 149 性证明。但在细胞中，它只提供了制造蛋白质的程序。从没有人想到DNA也能解决和计算机领域里一样的问题，直到1994年伦纳德·艾德曼才提出这个观点。艾德曼意识到，基因编码就像计算机科学中的二进制编码一样，也可以编码数学问题。他展示了如何利用生物技术操纵重组DNA来生成一个问题的各种可行解答。之后再使用DNA序列分析技术进行筛选，在所有可行解答中鉴别出正确的解答。

对于特定类型的数学问题，并没有用计算机找出正确解答的简便办法。计算机必须逐一检验所有可能的选项，从中挑出最优的一个。如果可行解答的数量太大，进行搜索就需要非常长的时间。这些都是用传统计算机最难以解决的一类问题。一个经典的例子是“旅行推销员”问题，问题内容是：给定大量的空间中的点（“城市”），求出能途经所有点各一次的最短路径。

艾德曼展示了，通过把DNA短片段打乱并拼接，问题的所有可行解答都可以在试管中编码成单链DNA分子。解答的数量可能非常巨大，但也比不上试管中的分子数量大。而且，所有可行解答都是同时产生并检验的，所以原则上DNA计算能够迅速地找出“最优”的解答。

无论DNA计算是否能够用于实践，它都有着寓言般的强大魅力。它把“生命分子基础的根源在于对信息的操作”这种观点演绎到了极致。人们常说，每个年代的人都倾向于用当时最先进技术所衍生出的模型来阐释世界，那么，面对那个困扰了霍尔丹、薛定谔和无数先贤的历史悠久的问题，或许身处信息时代的我们也应当谨慎，不要对这个（片面的）答案过于武断。或许更重要的是，我们应该把这看作一种启示，展示的是这个无限动态的、相互作用的、我们未曾看见又常常疏于讴歌的分子世界。

150

索 引

（条目后的数字为原书页码，见本书边码）

A

acetylcholine 乙酰胆碱 125, 126, 129

acetyl coenzyme A 乙酰辅酶 A 80

actin 肌动蛋白 103—106

action potential 动作电位 124, 126

adenine 腺嘌呤 135

adenylate cyclase 腺苷酸环化酶 119, 120

Adleman, Leonard 伦纳德 · 艾德曼 150

adrenal gland 肾上腺 117

adrenaline 肾上腺素 117

alizarin 茜素红 27, 28

allosteric effects 变构效应 85

alpha helix α螺旋 39, 41

amino acids 氨基酸 40, 46, 53, 125

amylases 淀粉酶 77

analgesics 镇痛药 127

anandamide 内源性大麻素 127

aniline dyes 苯胺染料 26

antibiotics 抗生素 28, 35

antidiuretic hormone 抗利尿激素 116

aramid fibres 芳纶纤维 61

aromatic compounds 芳香族化合物 25, 27

arsenic 砷 6

aspirin (acetylsalicylic acid) 阿司匹林（乙酰水杨酸）28, 127

Atkins, Peter 彼得 · 阿特金斯 74

atomic force microscope 原子力显微镜 107

atomic theory 原子理论 3, 6

atoms 原子 2, 3, 4, 9, 11, 16, 20

  atomic symbols 原子符号 21

ATP (adenosine triphosphate) 三磷酸腺苷 76, 79, 81—83, 87—90, 99, 101, 103, 104, 110, 111, 119, 123

ATP synthase ATP 合成酶 39, 83, 89, 90, 111, 112

autotrophs 自养生物 86

Aviram, Ari 阿里 · 艾维瑞姆 146

axon 轴突 123—124

axoneme 轴丝 99, 100

B

bacteria 细菌 42, 63, 84

Balzani, Vincenzo 温琴佐 · 巴尔扎尼 148

benzene 苯 25, 26

Berzelius, Jons Jacob 约恩斯 · 雅各布 · 贝采里乌斯 35, 113

beta sheet β折叠 40, 57, 58, 59, 60, 63

binary data 二进制数据 146—150

bioenergetics 生物能学 70—90, 115

biology 生物学 7, 12, 35—37, 70, 96, 97, 112, 114, 149

  molecular 分子的 33—49, 113

biotechnology 生物技术 62, 63, 141, 142

block copolymers 嵌段共聚物 62

blood 血液 77, 78, 84—86, 113, 115, 117

bonding, in molecules 分子中的化学键 16, 17

non-covalent 非共价键 129, 130
bone 骨骼 52, 54
Boyer, Paul 保罗·博耶 83
Boyle, Robert 罗伯特·玻意耳 6
brewing 酿酒 84
buckminsterfullerene 巴克敏斯特富勒烯，见 C_{60}

C

C_{60} C_{60}分子 66
calmodulin 钙调蛋白 121
Calvin, Melvin 梅尔文·卡尔文 89
Calvin-Benson cycle 卡尔文-本森循环 88—89
cancer 癌症 29, 66, 118, 140
cannabinoids 大麻素 127
carbohydrates 糖类 75, 76
carbon 碳 20, 24, 25, 66—69
carbon cycle 碳循环 86—87
carbon dioxide 二氧化碳 70, 80, 84, 86, 87—89
carbon fibres 碳纤维 68
carbon monoxide 一氧化碳 82, 128
carbon nanotubes 碳纳米管 66—69
catalysis 催化 37, 39, 60, 128, 131
cell division 细胞分裂 65, 100, 110, 139
cells 细胞 20, 36, 37, 38, 43, 44, 45, 99, 113
celluloid 赛璐珞 91
cellulose 纤维素 53, 77, 91
ceramics 陶瓷 51
chemical formula 化学式 22
chemistry 化学 3, 8, 95
history of 化学史 4—8
in industry 化学工业 6, 9, 26—28, 114
in medicine 医药化学 28, 29, 52, 113, 114
of 'informed matter' "信息物质"的 134
synthetic 合成的 23—32
chemotherapy 化学疗法 28
chlorophyll 叶绿素 89
chloroplasts 叶绿体 87—90
chromosome 染色体 42, 65, 100, 133, 141
cilia 纤毛 64, 98
citric acid cycle 柠檬酸循环 73—83, 88
Clarke, Arthur C. 亚瑟·C. 克拉克 50
coal-tar products 煤焦油产品 25—27
collagen 胶原蛋白 53—55, 63
combustion 燃烧 72, 75, 92
computing 计算 145
molecular 分子计算 134, 147—151
DNA DNA计算 149, 151
conducting polymers 导电聚合物 146
contraceptive pill 避孕药 47
Corey, Elias J. 伊莱亚斯·J. 科里 30
cornea 角膜 55
cortisol 皮质醇 117
covalent bond 共价键 17
Crick, Francis 弗朗西斯·克里克 44
crown ethers 冠醚 129—130
curare 箭毒 126
cyclic AMP 环状AMP 119—122
cyclonite 黑索今炸药 92
cytochrome enzymes 细胞色素酶 81—82
cytosine 胞嘧啶 135—138

D

Dalton, John 约翰·道尔顿 6, 21
Darwin, Charles 查尔斯·达尔文 35
Democritus 德谟克利特 6
denaturation 蛋白质变性 41
de Silva, A. Prasanna A. 普拉桑纳·德·席尔瓦 148
diabetes 糖尿病 79
diamond 钻石 68, 69
Dickinson, Emily 艾米莉·狄金森 33
digestion 消化 75—79
DNA (deoxyribonucleic acid) 脱氧核糖核酸 20, 41—48, 62, 133, 135—145
  double helix 双螺旋 136, 137
  replication 复制 137, 139, 140
  in supramolecular chemistry 超分子化学中的 142—145
DNA computing DNA 计算 149, 151
DNA polymerase DNA聚合酶 23, 137, 141
dopamine 多巴胺 125
Drexler, K. Eric K. 埃里克·德雷克斯勒 107
drugs 药物 28, 113, 126—128
dye industry 染料工业 26—28
dynamite 甘油炸药 90, 91
dynein 动力蛋白 99—101

E

Ehrlich, Paul 保罗·埃利克 28
Eiffel Tower 埃菲尔铁塔 53, 54
elastin 弹性蛋白 39, 63
electron microscope 电子显微镜 20, 67
electron transport 电子的输送 82, 87—88
electrons 电子 17, 81
elements 元素 4, 6
Empedocles 恩培多克勒 6
endoplasmic reticulum 内质网 42, 101
endorphins 内啡肽 127
entropy 熵 73—74, 132
enzymes 酶 22, 38, 47, 74, 77—79, 81, 85, 89, 119—121, 129, 137, 140, 141
epinephrine 肾上腺素 78
evolution 进化 35, 45, 140
exons 外显子 141
explosives 炸药 90—92

F

Faraday, Michael 迈克尔·法拉第 72
fats 脂肪 75
Feringa, Ben 本·费林加 107
fermentation 发酵 63, 84
Feynman, Richard 理查德·费曼 94—98
fibrous materials 纤维材料 31, 53—61, 63, 66, 68, 69
fingernails 指甲 52, 56
Fischer, Emil 埃米尔·菲舍尔 128
Fleming, Alexander 亚历山大·弗莱明 28
follicle-stimulating hormone (FSH) 卵泡刺激素 117
fullerenes 富勒烯 66, 67

G

GABA (gamma-aminobutyric acid)

γ-氨基丁酸 126
gametes 配子 139
GDP (guanosine diphosphate) 鸟苷二磷酸 119, 120
gelignite 炸胶 91
genes 基因 12, 34, 42, 44, 45, 53, 62, 134—136, 139—142
genetic code 遗传密码 135
genetic engineering 基因工程 62, 63, 140, 141, 144
genetics 遗传学 42—46, 133—142
genome 基因组 133, 141
Gilman, Alfred 阿尔弗雷德·吉尔曼 121
glands 腺体 115—117
glucagon 胰高血糖素 78, 115
glucose 葡萄糖 76, 78, 79, 87, 89, 90, 115
glycine 甘氨酸 57, 125, 126
glycogen 糖原 78, 117
glycolysis 糖酵解 79, 82, 83, 87
Goodyear, Charles 查尔斯·固特异 59
G proteins G 蛋白 119—122, 125, 131
Graebe, Carl 卡尔·格雷贝 27
graphite 石墨 66
Gravity's Rainbow《万有引力之虹》8, 10
growth hormone 生长激素 116
GTP (guanosine triphosphate) 鸟苷三磷酸 76, 119, 120
guanine 鸟嘌呤 135—138
gun cotton 火药棉 91
gunpowder 黑火药 90

H

haeme group 血红素基团 85, 86
haemocyanin 血蓝蛋白 86
haemerythrin 蚯蚓血红蛋白 86
haemoglobin 血红蛋白 39, 85, 128
hair 毛发 52, 55
Haldane, J. B. S. J.B.S. 霍尔丹 33, 36, 37, 44, 151
heat 热量 73
Heath, James 詹姆斯·希斯 146—147
hemlock 毒芹 147
hierarchical structure 层级结构 53—55, 58, 101
HMX (high-melting explosive) 高熔点炸药 92
Hoffmann, Roald 罗阿尔德·霍夫曼 8
Hofmann, August Wilhelm 奥古斯特·威廉·霍夫曼 25, 26
Holton, Robert 罗伯特·霍尔顿 30
Hopfield, John 约翰·霍普菲尔德 133
hormone replacement therapy 激素替代疗法 117, 118
hormones 激素 78, 115—122
horn 角 53, 55
human genome project 人类基因组计划 133
Huxley, Aldous 阿道斯·赫胥黎 91
Huxley, Andrew 安德鲁·赫胥黎 102
Huxley, Hugh 休·赫胥黎 102
hydrocarbons 烃 59, 60, 90
hydrogen 氢 20
hydrogen bond 氢键 41, 57, 58, 136, 137
hydrolysis 水解 76

hypothalamus 下丘脑 115, 116

I

IG Farben 法本公司 8
Iijima, Sumio 饭岛澄男 66—67
information, molecular 分子信息 12, 132—151
inorganic chemistry 无机化学 32
insulin 胰岛素 78, 79, 115
intestine 肠 77
introns 内含子 140
ion channels 离子通道 83, 122, 123—126
ionophores 离子载体 129
iron 铁 9, 85
isomers 同分异构体 22, 108

J

Jacob, François 弗朗索瓦·雅各布 43
'junk' DNA "冗余" DNA 47, 138—141

K

Kekulé, August Friedrich von 奥古斯特·弗雷德里克·冯·克库勒 26
Kelly, Ross 罗斯·凯利 108, 109
keratin 角蛋白 55
Kevlar 凯夫拉 61, 68
Kiedrowski, Günter von 君特·冯·凯德罗夫斯基 138
kinesin 驱动蛋白 101, 109, 111
Kinosita, Kazuhiko 木下一彦 106
Koert, Ulrich 乌尔里克·薛尔特 130

L

lactate 乳酸 79
Lavoisier, Antoine 安托万·拉瓦锡 6
Lehn, Jean-Marie 让-玛丽·莱恩 129, 134
Leibler, Stanislas 斯坦尼斯拉斯·莱布勒 110
Leucippus 留基伯 6
Levi, Primo 普里莫·莱维 4, 6, 8, 14, 15, 16, 24, 29, 95, 96, 105
Liebermann, Carl 卡尔·利伯曼 27
light 光 17, 18, 86
lipids 脂类 90, 133
liposomes 脂质体 90
living polymerization 活性聚合 69
'lock-and-key' mechanism "锁钥"机理 118, 128
logic gates 逻辑门 181
LSD (lysergic acid diethylamide) 麦角酰二乙胺 147
luteinising hormone 黄体生成素 117
lysozyme 溶菌酶 22

M

McClenaghan, Nathan 内森·麦克莱纳根 148
materials, molecular 分子材料 50—69
mauve (coal-tar dye) 苯胺紫(煤焦油染料) 26
Mendel, Gregor Johann 格雷戈尔·约翰·孟德尔 44
Mendeleev, Dmitri 德米特里·门捷

列夫 6
mescaline 墨司卡林 127
metabolism 代谢作用 70, 71, 76—84, 115
aerobic 有氧的 79—84
anaerobic 无氧的 79, 83, 84
micromachines 微机械 95, 97
microscopy 显微镜学 17—20
microtubules 微管 63—65, 99, 111
Mirkin, Chad 查德・米尔金 144
mitochondrial DNA 线粒体 DNA 45
mitochondrion 线粒体 45, 80—83, 85
mitotic spindle 纺锤体 65
molecular electronics 分子电子学 32, 144—148
molecular formula 分子式，见 chemical formula
molecular logic 分子逻辑 145—151
molecular motors 分子马达 97—104, 107—112
molecular phylogeny 分子种系发生学 45
molecular recognition 分子识别 118, 128—130, 136, 142—145
molecular sensors 分子传感器 130, 131
molecular switches 分子开关 146—149
molecules 分子：
definition 定义 9—11
in life 生命分子 20, 23, 33—49, 70
organic 有机的 24
structure 结构 12, 14—23
Monkey's Wrench, The《猴子的扳手》14, 96
Monod, Jacques 雅克・莫诺 43
Montemagno, Carlo 卡罗・蒙泰马尼奥 111
Moore, Gordon 戈登・摩尔 145
Moore's law 摩尔定律 145
morphine 吗啡 127
motor neurons 运动神经元 103
motor proteins 马达蛋白 97—105, 110—112, 145
Müller, Achim 阿希姆・缪勒 31, 32
muscle 肌肉 52, 101—106
mutation, genetic 基因突变 45, 140
myofibril 肌原纤维 101, 102
myoglobin 肌红蛋白 39, 85
myosin 肌球蛋白 103—106, 109

N

NAD (nicotinamide adenine dinucleotide) 烟酰胺腺嘌呤二核苷酸 81, 84, 88
NADP (NAD phosphate) 磷酸 NAD 88—89
nanocrystals 纳米晶体 144, 145
nanotechnology 纳米技术 98, 107, 110—112, 114, 142—145
natural products 天然产物 27
nerve signalling 神经信号 115, 122—128, 133
neuromuscular junction 神经肌肉接点 103, 125, 126, 127
neurons 神经元 122—124, 126
neurotoxins 神经毒素 126
neurotransmitters 神经递质 124—128
Nicolaou, K. C. K.C. 尼古拉奥 30
nicotine 尼古丁 126

nitric oxide 一氧化氮 128
nitrocellulose 硝化纤维 91
nitroglycerin 硝化甘油 91, 128
Nobel, Alfred 阿尔弗雷德·诺贝尔 89, 90, 92
noradrenaline 去甲肾上腺素 117
nucleotide 核苷酸 42, 134, 136
nucleus 核:
  of atoms 原子核 16
  of cells 细胞核 42

O

O Brien, Flann 弗兰·奥布赖恩 1, 3, 9, 20, 70
octanitrocubane 八硝基立方烷 92
oestrogen 雌激素 117
oleamide 油酰胺 127
olfaction 嗅觉 121, 122, 128
operon 操纵子 43
opium 鸦片 127
optical tweezers 光镊 105—106
organelles 细胞器 100
organic chemistry 有机化学 25—31, 35
Orgel, Leslie 莱斯利·奥格尔 137
origin of life 生命起源 46—48
oxygen 氧 20, 75, 79, 81, 82, 84, 85—86, 89, 117

P

Paclitaxel 紫杉醇，见 taxol
Pauli, Wolfgang 沃尔夫冈·泡利 6
peptide bond 肽键 40
Periodic Table 周期表 3, 5, 6, 8
Periodic Table, The (Primo Levi)《元素周期表》(普里莫·莱维) 3, 4
Perkin, William 威廉·珀金 25—27
Plato 柏拉图 6, 7, 15
pepsin 胃蛋白酶 77
penicillin 青霉素 28
phosphorylation 磷酸化 119
photosynthesis 光合作用 86—89
  artificial 人工的 90
photosystem 光系统 88
physics 物理学 7, 12
pituitary gland 垂体 115—117
plants 植物 82—86
plastics 塑料 51, 53, 59, 60, 91
polyethylene 聚乙烯 59, 61
polymerase chain reaction (PCR) 聚合酶链式反应 141
polymers 聚合物 10, 41, 53, 59—63, 69, 91
polypeptide 多肽 41
porphyrins 卟啉 85
potassium channel 钾通道 123—126, 129
progesterone 孕酮 117
proof-reading enzymes 校对酶 140
prostaglandins 前列腺素 127
protein folding 蛋白质折叠 41, 42
protein kinases 蛋白激酶 119
proteins 蛋白质 12, 20, 23, 37—42, 43, 46, 52, 75, 101, 133, 136, 139
  structural 结构蛋白，见 structural proteins
  structure 结构 38—42, 135
Pynchon, Thomas 托马斯·品钦 8
pyruvate 丙酮酸 79, 80, 84

Q

quantum theory 量子理论 6, 20
quinine 奎宁 25, 26

R

Ratner, Mark 马克·拉特纳 146
rayon 人造丝 91
receptor proteins 受体蛋白 118—120, 121, 125, 126, 131, 148
red blood cells 血红细胞 85
Reed, Mark 马克·里德 148
regulatory genes 调节基因 43
replication, molecular 分子复制 48, 49, 138
respiration 呼吸作用 80, 81, 84—86
retrosynthetic analysis 逆向合成分析 30
ribosome 核糖体 23
ribozymes 核酶 47, 48
RNA (ribonucleic acid) 核糖核酸 42, 47, 137
RNA editing RNA 编辑 47, 140
RNA world RNA 世界 46—47
rocket wire 火箭丝 61
Rodbell, Martin 马丁·罗德贝尔 121
rotaxane 轮烷 147—149
rubber 橡胶 59

S

salt (sodium chloride) 食盐(氯化钠) 11
Salvarsan 撒尔佛散 28
sarcomere 肌节 102
sarcoplasmic reticulum 肌浆网 103
scanning probe microscope 扫描探针显微镜 17, 106
scanning tunneling microscope 扫描隧道显微镜 19
Schonbein, Christian 克里斯蒂安·舍恩拜因 91
Schrodinger, Erwin 埃尔温·薛定谔 74, 122, 151
Second Law of Thermodynamics 热力学第二定律 73, 74, 98
Seeman, Nadrian 纳德里安·希曼 143—144
self-assembly 自组装 142—143
self-organization 自组织 132
sequencc 序列:
DNA 脱氧核糖核酸 41, 135
protein 蛋白质 41, 57, 58, 62—63, 136
serotonin 血清素 125, 127
sex glands 性腺 117
shear-induced alignment 剪切诱导对齐 61
signal transduction 信号转导 118—121, 128, 130—131
silk 丝 51, 56—62, 68
spinning 吐丝 58, 59, 61, 62
silkworm 蚕 58
skin 皮肤 52, 55, 56
Skou, Jens 延斯·斯科 83
Smalley, Richard 理查德·斯莫利 68
smart materials 智能材料 52
Sobrero, Ascanio 阿斯卡尼奥·索布

雷罗 91
sodium 钠 5
sodium channel 钠通道 122, 123—125, 126
Space Elevator 空间电梯 50, 51, 68, 69
space tethers 空间绳索 51
starch 淀粉 77
stereoscopic images 立体影像 22
steroids 类固醇 117
Stoddart, Fraser 弗雷泽·斯托达特 147
structural genes 结构基因 43
structual proteins 结构蛋白 38, 52—58, 60—63
strychnine 士的宁 127
sugar 糖，见 glucose
supramolecular chemistry 超分子化学 114, 128—131, 142—145, 147—149
symbiosis 共生 84
synapses 突触 123
synthesis 合成 10, 23—32, 60, 62
rational 定向的 27, 29—32

T

taste 味道 122
taxol 紫杉醇 29—32, 66
telomeres 端粒 141
testosterone 睾酮 117
thermodynamics 热力学 73—75
thylakoid membranes 类囊体膜 87—89
thymine 胸腺嘧啶 135—137
thyroid gland 甲状腺 115, 116
thyroid-stimulating hormone 促甲状腺激素 116
thyroxine 甲状腺素 115
TNT (trinitrotoluene) 三硝基甲苯 91
tooth 牙齿 52, 55
Tour, James 詹姆斯·图尔 148
transcription 转录 42, 47, 138, 139
translation 翻译 42, 47, 138—141
transposons 转位子 141
travelling salesman problem 旅行推销员问题 150
triiodothyronine 三碘甲腺原氨酸 115
tropomyosin 原肌球蛋白 103, 104
troponin 肌钙蛋白 103, 104
tryptophan 色氨酸 39
tubulin 微管蛋白 64

U

ubiquitin 泛素 39
uranium 铀 6, 8

V

valence 化合价 20
valinomycin 缬氨霉素 129
Viagra 伟哥 128
viruses 病毒 37
vitalism 生命活力论 35, 74
vitamin C 维生素 C 53
Vogel, Viola 维奥拉·沃格尔 111

W

Walker, John 约翰·沃克 83
water 水 11, 14, 16, 17, 89

X

Z

Philip Ball

MOLECULES

A Very Short Introduction

Contents

Preface

When Alexander Findlay wrote *Chemistry in the Service of Man* in 1916, there was an urgent need to advertise the benefits that chemistry had brought the world. Nine decades later, those writing about chemistry might hope to have been relieved of that burden. But it is not so. In spite of the single most dramatic contribution of chemical art to society – the increase in life span owing to chemotherapeutic health care – Findlay's words still have a familiar ring:

> The people as a whole, being ignorant of science, have mistrusted and looked askance at those who alone could enlarge the scope of their industries and increase the efficiency of their labours.

This same sternness of tone is often not far beneath the surface of efforts today by the chemical industry and its advocates to defend itself against public disdain and censure. One of the problems is that, while the good is taken for granted almost as soon as it is brought to market, the bad sticks in the mind for years. And there is no denying that the attempts by chemicals companies and governments to shirk responsibility for tragedies such as thalidomide and Bhopal, or near-catastrophes such as ozone depletion, have left them with severely diminished credibility to plead their case.

Thus we face the twenty-first century with a pervasive feeling that 'chemical' or 'synthetic' is bad, and 'natural' is good.

The traditional remedy is to list all the good things that chemistry has given us. This list is indeed long, and those who would demonize industrial chemistry probably enjoy many of its products. But I believe that 'chemistry in the service of man' is no longer what we need. For one thing, it perpetuates the impression of a monolithic scientific and technological enterprise universally committed to advancing its own cause. To outsiders, any culture looks monolithic and therefore potentially threatening. It will be a good day when there is more public recognition of how chemists argue furiously with one another about whether this or that product should be banned or restricted, or of the fact that some chemists work in military establishments while others join the blockade outside the gates. Maybe then we will start to see science as a human activity.

But, secondly, chemistry is not simply a thing to be tamed and commandeered into service. It is also what *makes* a man or woman, and the rest of nature too. The negative connotations of 'chemical' and 'synthetic' are hard now to shrug off; but 'molecules' have not yet acquired such colours. And it is by understanding our own molecular nature that we can perhaps begin to appreciate what chemistry has to offer, as well as perceiving why it is that some substances (natural and artificial) poison us and some cure us.

This is why I risk disapproval from some chemists by writing a guide to molecules that focuses to a large extent on the molecules of life – on biochemistry. What I have tried to show is that the molecular processes that govern our own bodies are not so different from those that chemists – I would prefer to say molecular scientists – are seeking to create. Indeed, the boundaries are becoming blurred: we are already using natural molecules in technology, as well as using synthetic molecules to preserve what we deem 'natural'.

In trying to tell these molecular tales, I have benefited greatly from the expert advice of Craig Beeson, Paul Calvert, Joe Howard, Eric Kool, Tom Moore, and Jonathan Scholey, to whom I extend my sincere thanks.

This book began its life as it will end it: as a contribution to OUP's Very Short Introduction series. I am very grateful to Shelley Cox for having sufficient belief in the text to offer it, for a time, an independent life.

Philip Ball

London
January 2001

Acknowledgements

The author and publisher gratefully acknowledge permission to use the following copyright material:

Extract from *The Dalkey Archive* by Flann O'Brien (copyright © The Estate of the Late Brian O'Nolan), reproduced by permission of A. M. Heath & Company Ltd. on behalf of the Estate.

Extract from *The Periodic Table* by Primo Levi, translated by Raymond Rosenthal, copyright © 1984 by Schocken Books, a division of Random House, Inc. Used by permission of Schocken Books, a division of Random House, Inc.

Extract from *Gravity's Rainbow* by Thomas Pynchon, copyright © 1973 by Thomas Pynchon. Reprinted by permission of Melanie Jackson Agency, L.L.C.

List of illustrations

Except where indicated, drawings are by the author

The publisher and the author apologize for any errors or omissions in the above list. If contacted they will be pleased to rectify these at the earliest opportunity.

Chapter 1

Engineers of the invisible: making molecules

> The sergeant beckoned the waitress, ordered a barley wine for himself and a small bottle of 'that' for his friend. Then he leaned forward confidentially.
>
> – Did you ever discover or hear tell of mollycules? he asked.
> – I did of course.
> – Would it surprise or collapse you to know that the Mollycule Theory is at work in the Parish of Dalkey?
> – Well . . . yes and no.
> – It is doing terrible destruction, he continued, the half of the people is suffering from it, it is worse than the smallpox.
> – Could it not be taken in hand by the Dispensary Doctor or the National Teachers, or do you think it is a matter for the head of the family?
> – The lock, stock and barrel of it all, he replied almost fiercely, is the County Council.
> – It seems a complicated thing all right.

The shortest of short introductions to molecules has already been written, and is far more witty than mine. Flann O'Brien was a man who liked to serve up his erudition over a pint of Guinness, as though he were discussing the potato crop or the terrible state of the roads out of Dublin. We can benefit from some more of the wisdom that Sergeant Fottrell is sharing with Mick in the Metropole Hotel, on Dublin's main street:

– Did you ever study the Mollycule Theory when you were a lad? he asked. Mick said no, not in any detail.

– That is a very serious defalcation and an abstruse exacerbation, he said severely, but I'll tell you the size of it. Everything is composed of small mollycules of itself, and they are flying around in concentric circles and arcs and segments and innumerable various other routes too numerous to mention collectively, never standing still or resting but spinning away and darting hither and thither and back again, all the time on the go. Do you follow me intelligently? Mollycules?

– I think I do.

– They are as lively as twenty punky leprechauns doing a jig on the top of a flat tombstone. Now take a sheep. What is a sheep but only millions of little bits of sheepness whirling around doing intricate convulsions inside the baste.

What is a sheep? This simple question is (under many guises) more than enough to have kept scientists occupied for hundreds of years, and will continue to do so for many years to come. The science of molecules gives an answer embedded in a hierarchy of answers. It is concerned with the 'millions of little bits of sheepness', which are called molecules. A sheep is a blend of many kinds of molecule – tens of thousands of different varieties. Many of them appear not only in sheep but in humans, in the grass, in the skies and oceans.

But science, seeking deeper levels of understanding, does not leave things there. Are not a sheep's molecules made of atoms, and are not atoms made of subatomic particles such as electrons and protons, and are not those made of sub-subatomic particles such as quarks and gluons, and who is to say what *they* contain within their absurdly tiny boundaries?

– Mollycules is a very intricate theorem and can be worked out with algebra but you would want to take it by degrees with

> rulers and cosines and familiar other instruments and then at the wind-up not believe what you had proved at all. If that happened you would have to go back over it till you got a place where you could believe your own facts and figures as exactly delineated from Hall and Knight's Algebra and then go on again from that particular place till you had the whole pancake properly believed and not have bits of it half-believed or a doubt in your head hurting you like when you lose the stud of your shirt in the middle of the bed.
>
> – Very true, Mick decided to say.

It is indeed an intricate business to work out what molecules are, if you want to begin on a lower (we should perhaps say deeper) rung of the ladder of science and climb upwards. That is necessary if one wishes fully to understand why molecules behave the way they do, and in consequence why matter – why a sheep or a rock or a pane of window glass – displays its characteristic gamut of properties. But many scientists who work with molecules do not need to bother with all the algebra, for its implications can be generally boiled down to rules of thumb about how molecules interact with one another. The chemical industry was a thriving enterprise before chemistry found its mathematics. Which is a way of saying that molecules need not, after all, make your head hurt.

Leaving the table

It is curious that, when Flann O'Brien reworked the conversation between Sergeant Fottrell and Mick from *The Dalkey Archive* into his most famous novel *The Third Policeman*, published after his death in 1966, he systematically replaced the 'Mollycule Theory' with the 'Atomic Theory'. Here then is the very item, the ambiguity about what things are made from. Is it atoms or molecules? Chemists give out mixed messages. Their iconic cryptogram is the Periodic Table, a list of the ninety-two natural

Elements: Primo Levi's *The Periodic Table*

There are the so-called inert gases in the air we breathe. They bear curious Greek names of erudite derivation which mean 'the New', 'the Hidden', 'the Inactive', and 'the Alien'. They are indeed so inert, so satisfied with their condition, that they do not interfere in any chemical reaction, do not combine with any other element, and for precisely this reason have gone undetected for centuries. As late as 1962 a diligent chemist after long and ingenious efforts succeeded in forcing the Alien (xenon) to combine fleetingly with extremely avid and lively fluorine, and the feat seemed so extraordinary that he was given a Nobel prize . . .

Sodium is a degenerated metal: it is indeed a metal only in the chemical significance of the word, certainly not in that of everyday language. It is neither rigid nor elastic; rather it is soft like wax; it is not shiny or, better, it is shiny only if preserved with maniacal care, since otherwise it reacts in a few instants with air, covering itself with an ugly rough rind: with even greater rapidity it reacts with water, in which it floats (a metal that floats!), dancing frenetically and developing hydrogen . . .

I weighed a gram of sugar in the platinum crucible (the apple of our eyes) to incinerate it on the flame: there rose in the lab's polluted air the domestic and childish smell of burnt sugar, but immediately afterward the flame turned livid and there was a much different smell, metallic, garlicky, inorganic, indeed contra-organic: a chemist without a nose is in for trouble. At this point it is hard to make a mistake: filter the solution, acidify it, take the Kipp, let hydrogen sulphide

bubble through. And here is the yellow precipitate of sulphide, it is arsenious anhydride – in short, arsenic, the Maculinum, the arsenic of Mithridates and Madame Bovary.

Primo Levi, *The Periodic Table* (1975)

elements (supplemented by some unstable, artificial ones) arranged in a pattern that helps chemists make sense of them. The most famous book 'about' chemistry is the one that Italian chemist and writer Primo Levi named after this tabulation of matter's building blocks, and it reinforces the impression that chemistry begins with this irregularly shaped grid of symbols. At school I was encouraged to learn mnemonics encoding the elements in the first two rows of the table, which are the most important. For undergraduate chemistry it was required that one could recite the whole thing from memory, to know that iridium lies at the foot of cobalt, that europium is sandwiched between samarium and gadolinium. Yet I doubt that I shall ever set eyes on samarium (although europium shines out at us redly from our television screens).

But chemistry is only incidentally about the properties of the elements, and the science of molecules can afford to ignore many if not most of them. The Periodic Table really belongs to that realm where chemistry becomes physics, where we must wheel out the algebra and the cosines to explain why atoms of the elements form the particular unions called molecules. The table is one of the most beautiful and profound discoveries of the nineteenth century, but, until quantum mechanics was invented by physicists in the twentieth century, one could look upon it only as a mysterious cipher, a kind of crib sheet that served as an empirical reminder that elements come in families whose members show similar proclivities.

Perhaps I am being too quick to dispense with the Periodic Table. At least, I should not do so without confessing to an agenda.

A conventional history of chemistry presents it as a quest to understand matter: to ask, what are things made of? This links chemistry with ancient Greek philosophy, with the attempts of Leucippus and his pupil Democritus to formulate an atomic theory of matter in the fifth and fourth centuries BC. It gives us a narrative that progresses from Empedocles' four elements – earth, air, fire, and water – through to Plato's marriage of elemental theory with atomism (Fig. 1), skirting cautiously around the medieval alchemists' belief in the transmutation of the elements and alighting gingerly on the phlogiston theory of the eighteenth century. We watch Robert Boyle redefine the idea of an element in 1661 (which, however, does not actually amount to much of a redefinition at all), we see antiquity's four-element scheme crumble before the discovery of new 'irreducible substances', and we see Antoine Lavoisier dismantle phlogiston and replace it with oxygen before losing his head under the guillotine's blade in 1794. John Dalton gives us the modern atomic theory in 1800, the list of elements expands enormously throughout that century, and then Dmitri Mendeleev arranges them into the twin-towered edifice of the Periodic Table. The gaps are gradually filled all the way up to uranium (itself known since 1789), and Wolfgang Pauli and the other quantum physicists explain the table's shape in the 1920s.

And so the task is at an end. According to science writer John Horgan in *The End of Science*, this meant that chemistry too was finished, once it had the quantum stamp of approval. The implication in several other recent books on the future of science is that the discipline, conspicuous by its absence, has been consumed from both ends. At the most fundamental level, it has become physics (including that immense but overlooked branch called condensed-matter physics, which ponders on how tangible matter

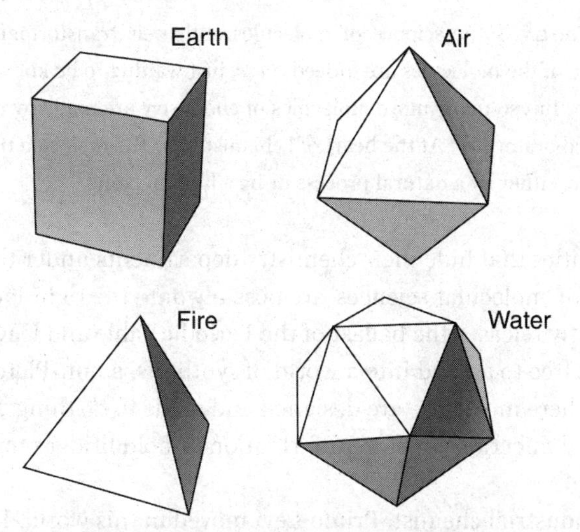

1. Plato's atoms. The Greek philosopher believed that the smallest particles of the four elements then thought to comprise everything had regular geometric shapes.

behaves). At the most complex level, it is now the domain of biologists, who have expanded their world to embrace the molecular mechanics of the cell.

But these academic turf wars conceal a far more interesting truth. It is a curious fact that many histories of science are written by physicists, who have a tendency to present science as a series of questions posed and then answered. It would be instructive to see the story told instead by an engineer, whose instinct might rather be to ask: what can we make? For, while some of our proto-chemists were wishing to dissect matter, whether physically or metaphysically, others were eagerly rearranging it. This is why the science of molecules is both a creative as well as an analytical pursuit. It has been, at various times in history, concerned with making ceramic pots, dyes and pigments, plastics and other synthetic materials, drugs, protective coatings, electronic components, machines the size of a bacterium. 'What is strange', says chemistry Nobel laureate Roald Hoffmann, 'is that chemists should accept the metaphor of discovery'. He goes on:

> Chemistry is the science of molecules and their transformations. Some of the molecules are indeed *there*, just waiting to be known by us . . . But so many more molecules of chemistry are made by us, in the laboratory . . . At the heart of [chemistry] is the molecule that is made, either by a natural process or by a human being.

Universities that hide their chemistry departments under the banner of 'molecular sciences' are possibly onto the right idea; for this gently releases the ballast of the Periodic Table and leaves the chemist free to ascend into a world of synthesis, a non-Platonic realm where molecules are designed and made to *do* things, such as cure viral infections or store information or hold bridges together.

As an industrial chemist, Primo Levi moved in this world. He felt a little apologetic about his molecular science: he called it 'a "low" chemistry, almost culinary'. But the power of 'low' chemistry is awesome. It shifts billions of dollars each year, it can make the sick healthy and the healthy sick. Hamburg and Dresden were laid waste by low chemistry, and chemical and biochemical warfare are now more feared in the West than nuclear war. Many people believe that the nuclear bomb was itself the product of physics, but writing $E = mc^2$ does not give you Hiroshima – only separating isotopically distinct molecules of uranium compounds did that. In *Gravity's Rainbow*, Thomas Pynchon has no doubt where the true power of science lies: the villain of his fantasy from the fag-end of the Second World War is not the Bomb but a new plastic, an 'aromatic heterocyclic polymer' called Imipolex G, developed in a conspiracy between Europe's giant chemicals companies IG Farben, Ciba, Geigy, Shell Oil, and ICI. The message is that 'stuff' speaks louder* than theories.

* After the end of the war, a group from the Allies assembled by Eisenhower claimed that 'Without IG [Farben]'s immense productive facilities, its far-reaching research, varied technical experience and overall concentration of economic power, Germany would not have been in the position to start its aggressive war in September 1939.' It was one of IG Farben's subsidiary companies, Degesch, that made the poison gas Zyklon B used in the concentration camps.

Does this mean that molecular science is bad? Of course not – it means that it is a craft full of possibilities. Wonderful, inspiring, inventive possibilities. Terrible, nightmarish possibilities. Mundane but useful things, bizarre things, hard-to-understand things. Molecular science might one day help people to grow a new liver. Raphael, Rubens, and Renoir painted with molecules. Molecules orchestrated the origin of life.

Synthesis: Thomas Pynchon's *Gravity's Rainbow*

The origins of Imipolex G are traceable back to early research done at du Pont. Plasticity has its grand traditions and main stream, which happens to flow by way of du Pont and their famous employee Carothers, known as the Great Synthesist. His classic study of large molecules spanned the decade of the twenties and brought us directly to nylon, which is not only a delight to the fetishist and a convenience to the armed insurgent, but was also, at the time and well within the System, an announcement of Plasticity's central canon: that chemists were no longer to be at the mercy of Nature. They could decide now what properties they wanted a molecule to have, and then go ahead and build it . . . A desired monomer of high molecular weight could be synthesized to order, bent into its heterocyclic ring, clasped, and strung in a chain along with the more 'natural' benzene or aromatic rings. Such chains would be known as 'aromatic heterocyclic polymers'. One hypothetical chain that Jamf came up with, just before the war, was later modified into Imipolex G.

Thomas Pynchon, *Gravity's Rainbow* (1973)

What are molecules?

So molecules make up everything there is? Not exactly. All matter (outside of some strange astrophysical environments) is made up of atoms; but atoms do not always organize themselves into molecules. (I cannot tell whether Flann O'Brien made the switch from 'mollycules' to atoms because he understood, or did not understand, this distinction.) Most atoms on their own are highly reactive – they have a predisposition to join up with other atoms. Molecules are collectives of atoms, firmly welded together into assemblies that may contain anything up to many millions of them.

But there is a further, subtle distinction to be made. Flann O'Brien's Sergeant Fottrell speaks of 'mollycules' of rock and of iron. Strictly speaking, there are no such things – at least, not in a block of everyday rock or iron. By molecules, we generally mean assemblies of a discrete, countable number of atoms. In the water molecule there are three atoms: two of hydrogen and one of oxygen. A glass of water contains trillions upon trillions of atoms, but a snapshot of the liquid – were it able to reveal such tiny details – would show that at any instant they are nearly all grouped into these three-atom molecules, like a gigantic crowd holding hands in families of three (Fig. 2*a*).

The atoms in iron, in contrast, do not cluster into discrete molecules. They stack together like cannonballs in a regular array that goes on and on, like a regimented battalion of soldiers. One cannot identify any grouping of the atoms – each is equidistant from its neighbours. The same is true of sodium and chlorine atoms in a crystal of sodium chloride (table salt (Fig. 2*b*)). When iron melts, the atoms simply jostle one another like an unruly crowd. But when ice melts, it is as if the hydrogen and oxygen atoms continue to hold hands in threes as the crystal falls apart. One would say that ice is a molecular solid – the atoms are clustered into molecules – whereas iron and rock salt are not.

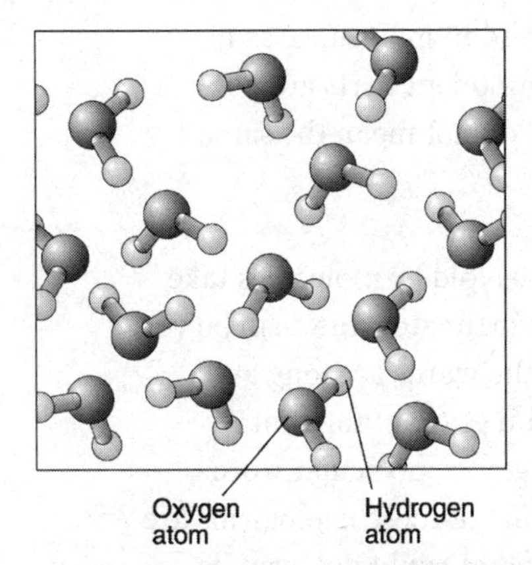

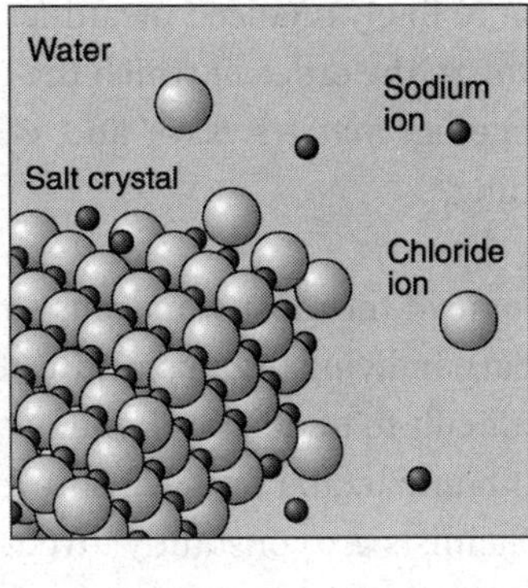

2. Water (*a*) is composed of discrete three-atom molecules, joined by strong chemical bonds. Salt (*b*), in contrast, is an assembly of charged atoms (ions) of sodium and chlorine, in which there are no discrete atomic groupings. When salt dissolves in water, the assembly merely falls apart ion by ion.

Some pure elements adopt molecular forms; others do not. As a rough rule of thumb, metals are non-molecular, like iron, whereas non-metals are molecular. Frozen nitrogen, for instance, consists of molecules containing two atoms each. In phosphorus the atoms form groups of four; in sulphur they can link into molecular rings of eight. It seems a little unfair that there is no way of knowing, simply by looking at a material, if its essential building blocks are atoms or molecular unions of atoms. But there is not. (It is not hard for scientists to find out, however.)

So 'molecule' is actually a rather fluid, loosely defined concept – essentially a question of scale. Why bother, then, to single out molecules at all, rather than simply talking about 'matter' in general? I would suggest the following reason: molecules are the smallest units of *meaning* in chemistry. It is through molecules, not atoms, that one can tell stories in the sub-microscopic world. They are the words; atoms are just the letters. Of course, sometimes a single letter constitutes a word. But most words are distinct aggregates of several letters arranged in a particular order. We often find that longer words convey subtler and more finely nuanced meanings. And in molecules, as in words, the order in which the component parts are put together matters: 'save' and 'vase' do not mean the same thing.

Some of the most wondrous stories told by molecules take place in living organisms. But unfortunately they can be very difficult to understand: many of the words are long and unfamiliar, and we have only a dim grasp of the syntax. Chemists are constantly inventing new molecular words, expanding the language – and some of these neologisms are rather witty. Some let us tell tales that could not even be formulated before the 'word' was invented. In other cases, a new 'word' allows us to say in a simple manner something that was previously conveyed in a roundabout way.

It is remarkable how nicely the linguistic metaphor fits the molecular world. We hear much today about the 'language of the genes', and I hope to show that this is just one of the tongues that molecules encode. Yet it is not merely a metaphor. There really is 'information' in molecules, just as there is in words, as I show in Chapter 7.

Moreover, using an information-based paradigm to describe molecular science is valuable in so far as it invites a responsive, dialogue-based description rather than the mechanical one that has been championed in former times. Cell biologists speak increasingly about protein molecules that 'talk to' one another; physicists interested in the science of matter speak of 'cooperative' and 'collective' behaviour. These are not woolly, romantic notions calculated to make science appear friendlier (although it will do no harm if they have that effect). Rather, they speak of the increasing awareness of the beautiful sophistication of molecular behaviour, which is generally gregarious and rarely linear.

It is with these thoughts in mind that I need to expand on the use of metaphor in molecular science. We cannot do without it, even at the level of one specialist speaking to another. This is true in many areas of science, but in chemistry more than most. Molecules are anthropomorphized mercilessly, and there need be no apology for that. They are unfamiliar things, these molecules, and we need to find ways of making them less so. The publishers of my book about water rightly insisted that ball-and-stick models of H_2O molecules were anathema to the non-chemist reader, guaranteed to ensure that the book stays on the shelf. Yet I could not explain water's strangeness without showing its molecular structure, and so I made the molecules into little demons (Fig. 3).

I hope this was harmless. But I was reminded of the dangers at a public lecture I attended recently on molecular replication. The first question from the floor was 'Are these molecules conscious?' Given that the speaker was talking about a synthetic molecular system that mimics (in a very crude way) some of the characteristics of living organisms, I suppose this was an understandable enquiry. I firmly believe the answer is 'no', if one wants to retain any meaningful working definition of the slippery concept of consciousness. But, once we start to anthropomorphize, we import a baggage of associations, for better or worse. Many people hate the concept of 'selfish genes' because it carries moral connotations. (Richard Dawkins calls it 'poetic science', and I can see what he means – but the poetry of the mechanism gets besmirched by the unpleasantness, as many see it, of the metaphor.) The idea that molecules 'cooperate' and 'communicate' is no basis for a philosophy of nature. But it is reason to suspect that, in molecular science at least, a linear, clockwork world view might in the end leave us like the ancient astronomers interpreting planetary motions from a geocentric perspective: trying to shoehorn the observations into a misconceived framework.

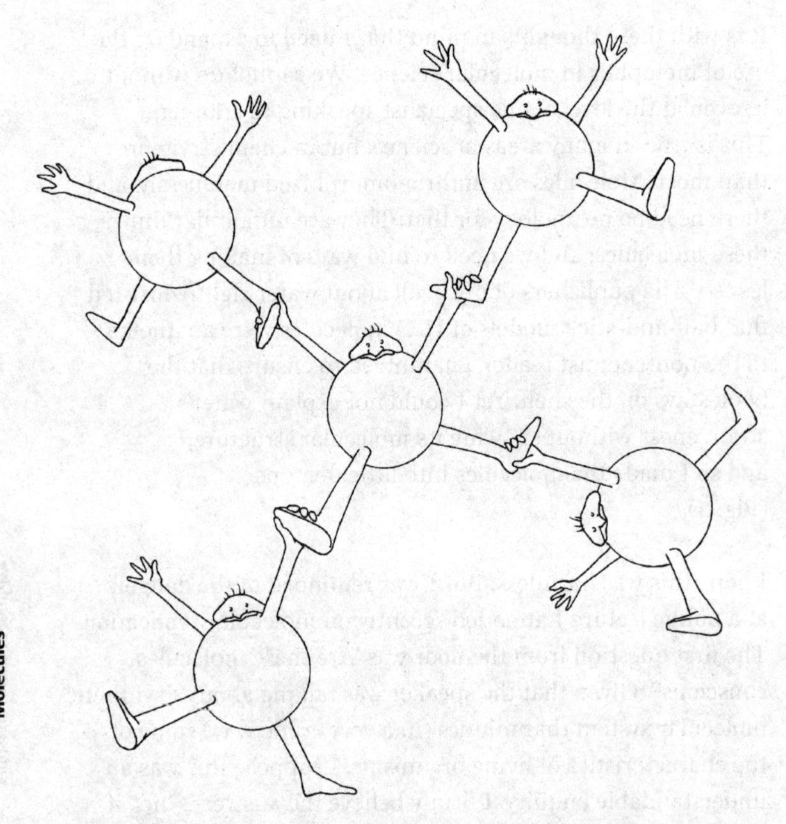

3. Making molecules anthropomorphic can help us visualize how they interact. Here I show the weak 'handclasps' that exist between water molecules.

Shape and size

Primo Levi's *The Monkey's Wrench* is one of the few novels I can think of that includes a drawing of a molecule (Fig. 4). It is a fearsomely complicated one, and I would never dream of showing it in a non-technical book about science if my intention was to be instructive.

But Levi gets away with it, because he does not want us to understand anything about the molecule, except for one thing: it

```
                                          H
                                          N–CO
                                         /    \
                                        CH2    N–
                                         \    /
      /                                   N–CH2
  CO–N                                   /
  /   \                      CH2–NHCO
–N     CH2                  /
  \   /               CO–N
 H2C–N                /   \
      \CO–NHCH2–N      CH2
                 \    /               /
                H2C–N             CO–N
                     \            /   \
                      CO–NHCH2–N       CH2
                                 \    /
                                 H2C–N
                                      \
                                       CO–NHCH2–
```

4. Primo Levi's molecule.

has a shape and structure. There are some kinds of hexagon in here, and some straight units linking them together. The narrator is talking to a construction worker named Faussone, a man who assembles girders into bridges. He says,

> the profession I studied in school and that has kept me alive so far is the profession of a chemist. I don't know if you have a clear idea of it, but it's a bit like yours; only we rig and dismantle very tiny constructions ... I've always been a rigger-chemist, one of those who make syntheses, who build structures to order, in other words.

We will encounter in these pages examples of molecules that can be regarded as miniature sculptures, containers, soccer balls, threads, rings, levers, and hooks, all made by sticking atoms together. Plato believed that atoms have the shapes of 'regular polyhedra': cubes, tetrahedrons, octahedrons, and so on. He was wrong;* but chemists can arrange atoms into molecules with these shapes.

* Actually, one can make a case that Plato was not far wrong at all. Atoms do link together in quite precise geometrical arrangements. Carbon atoms, for example, like to sit at the centre of a tetrahedron with four other atoms at the corners. This does not exactly make it the tetrahedral block that Plato envisaged for atoms of 'fire'; but it shows that Plato's geometric view of the microscopic world held a grain of truth.

So how big is this molecule that Levi's narrator draws for Faussone? Each one of those C's, N's, and so forth represents an atom, which is a truly tiny thing. Countless analogies struggle to convey the scale of atoms, but I am not sure that they serve to give an impression any more concrete than that these irreducible particles of the elements are very, very small indeed. For example, if a golf ball were blown up to the size of the Earth, its atoms would be about the size of the original golf gall. Ten million atoms of carbon side by side would make a row about a millimetre long.

A small molecule like water is just a few atoms' width in size, about three-tenths of a nanometre. (A nanometre is a millionth of a millimetre.) Primo Levi's molecule is several times bigger. (One cannot say exactly how many times, because what he drew was really just a fragment of a molecule, which continues to the right and the left of the page.)

One consequence of this scale is that things happen very fast in the molecular world. When we hear that molecules can rotate ten billion times a second, we imagine that they must be spinning at unimaginable speeds. But molecules are so small that, even if they travel at quite moderate speeds, they can cover molecular-scale distances in an instant. The atoms of an oxygen molecule need move only at a speed of about a metre per second to complete ten billion revolutions in a second.

What about the sticks that join the atoms together? In fact, they take up no space; they are just a convention to help us see what is going on in the diagram. Atoms that are bound together in molecules push right up against one another; in fact, they overlap, rather like two soap bubbles in contact. This is possible because atoms are not like hard billiard balls, but more like rubber balls. They have a centre that *is* dense and hard, called the nucleus, and this is about ten thousand times smaller than the atom itself – although it is where nearly all of the atom's mass is concentrated. The nucleus has a positive electrical charge. Surrounding it is a cloud of electrons, which are small, light subatomic particles with a negative charge. The electron clouds of two atoms can overlap without danger of electrons colliding, and the two atoms then share some of their electrons: the two clouds merge into one, encompassing both nuclei. When this happens, the two atoms are said to be linked by a *covalent bond*. The sticks in the molecular diagram on p. 15 represent covalent bonds, and they are just a way of helping us to see which atoms are bonded to which.

Here is one of the crucial considerations in talking about molecules, and it is one that complicates the whole attempt: there is no 'best' way of drawing them. One might say: well, never mind the schematic diagrams, why not show what they 'really' look like? But that does not help, because there is no way of taking a photograph of a molecule in the same way as we can photograph a cat or a tree. This is not a matter of technical limitations – it is not that we lack a microscope or a camera capable of resolving such small objects. The fact is the mechanics of seeing make it impossible to 'see' a molecule (or an atom, for that matter) 'as it really is'.

The reason is that we see with visible light, which is a wave-like radiation for which the wavelength – the distance between successive crests – varies from about 700 nanometres for red light to 400 nanometres for violet light. In other words, red light fits about 140,000 undulations into a centimetre. This wavelength is hundreds of times larger than a molecule. Roughly speaking, light cannot be focused to a point smaller than its wavelength, which means that objects smaller than that cannot be resolved.* No light-based microscope will ever show us a sharp image of a water molecule.

* I'm speaking here of conventional microscopy, where the light is focused by lenses. There are some new optical (light-based) microscopes that surpass this wavelength-limited resolution by getting the light source up close to the sample and shining it through a tiny aperture. This can increase the resolution to, so far, around a tenth of a wavelength.

I suspect that this is one reason why people find molecules hard to comprehend, and why diagrams like the one above are a good way to scare readers away from a science book. It seems absurd to be talking in a concrete manner about objects that are not only too tiny to see in practice but too tiny to see in principle. Things that cannot be seen acquire an aura of fantasy, as though they are just a convenient fiction.

Molecules are not a fiction, however, and we can prove not only that they are there but that they have definite shapes and sizes. Fig. 5 shows some portraits of molecules taken with a special kind of microscope that does not use light to form its images. Beside each snapshot I show a diagram of the molecule's structure. Well before this type of microscope was invented, the molecules were known to possess these structures; but no one had ever seen them directly. The images are pretty blurry – you would not be able to guess the exact shape of the molecules from these portraits alone. But the shapes seen in the microscope do match up in a convincing way with those expected.

How did we already know the shapes of these molecules before the pictures were taken? Some of the corroborating evidence is experimental. Even though molecules are too small to be resolved with visible light, they can be 'seen' with radiation of a wavelength comparable to their own size. Radiation with a wavelength of about a tenth of a nanometre corresponds to X-rays, and, by bouncing X-rays off crystals, it is possible to deduce where their constituent atoms are located. This means that, if a substance can be made in crystalline form, with all its molecules stacked together in an orderly manner, the technique called X-ray crystallography can reveal the structure of the molecules.

In principle we should be able to see individual molecules with an X-ray microscope that focuses X-rays just as we focus light in an optical microscope. In practice it is very hard to focus X-rays,

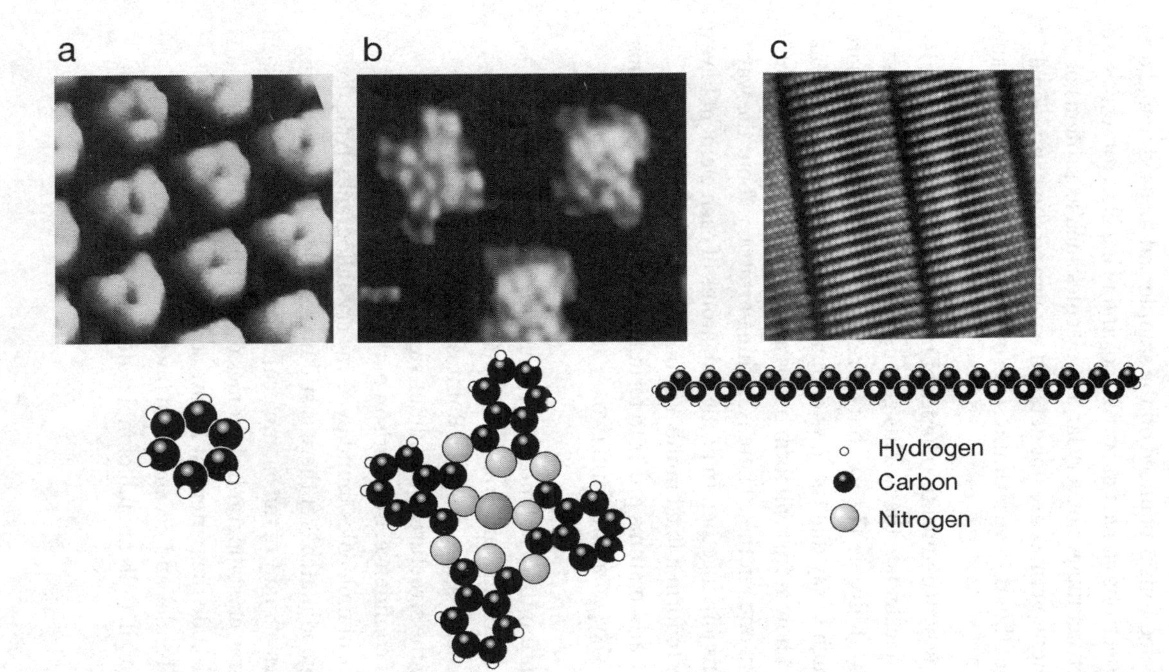

5. Molecules 'photographed' with the scanning tunnelling microscope, which is capable of resolving them individually. The STM is not (yet?) capable of showing sufficient detail to allow us to interpret the images without prior knowledge, however.

although scientists are on the verge of being able to do so. In the meantime we can get by with the electron microscope, in which a beam of electrons is bounced off the sample and focused to make an image. Electrons can act as waves too, and using electron waves we can construct images of very large molecules such as proteins or DNA. These pictures are not detailed enough to show individual atoms, but they do give an impression of the molecules' overall shape.

Another way to deduce the shapes of molecules is theoretical: it is possible to calculate them. This gets us into the 'algebra' of Flann O'Brien's 'Mollycule Theory', but there is no need to describe it here. It is enough to say that the laws of quantum mechanics* enable us to predict how atoms will form bonds and where they will then sit in relation to one another. There is nothing arbitrary about the way that atoms join together. In particular, atoms of each element have a tendency to form a fixed number of bonds, which is called its valence. Carbon atoms prefer to form four bonds, hydrogen atoms just one. Oxygen atoms form two.

The quantum theory of molecular structure is indeed 'a very intricate theorem', and even the best computers can solve the equations only approximately. But it is currently possible to calculate the structures of medium-sized molecules with a fair degree of confidence. Comparison of the predictions with the structures of molecules found by X-ray crystallography typically shows a good match. Yet there is still no reliable way of predicting the shapes of many of the big molecules found in living cells. In such cases, X-ray crystallography becomes difficult too, both because it is hard to decode the pattern of X-rays scattered from a crystal of these molecules and because in many cases they refuse to form crystals at all. Cells are full of molecules whose shapes we do not know.

* Quantum mechanics is a mathematical description of matter and its behaviours at very small scales, typically atomic dimensions. At this scale, matter can display wave-like properties.

This is a big hindrance to understanding how life's molecules do their jobs, because a molecule's shape is the key to its behaviour. To reverse a designer's motto, function follows form.

Molecular science is therefore a supremely visual science. Chemists have spent over two hundred years developing pictorial languages to describe their craft, with the result that they now have to be polylingual. There are many different ways to depict molecules, each developed to show the particular aspect that the illustrator wants to emphasize. The English chemist John Dalton began in 1800 by drawing molecules as collections of circular symbols depicting atoms, each embellished with shading or markings to identify the element concerned. It was transparent enough, once you knew the code (Fig. 6).

Very well – but not easy for the printer, who had to make up these symbols specially. A neat shorthand was to give each element a one- or two-letter symbol: C for carbon, O for oxygen, Ca for calcium, Fe for iron. (Classically minded even in the nineteenth century, chemists chose to register the metal as *ferrum*; gold and silver, *aurum* and *argentum*, became Au and Ag in the same way. 'Ir' is not iron but the element iridium. But at least the system was *meant* to be self-evident.)

Then carbon monoxide is simply CO. A multiplicity of atoms of the same element is denoted by a subscript, making the hydrogen molecule H_2.

Yet this scheme does not provide a unique inscription for every molecule. Dimethyl ether and ethanol are different substances with

6. Dalton's molecules.

different properties, yet both can be represented by the *chemical formula* C_2H_6O. We are back to the lexicological problem: the meaning of a word is determined not only by the letters it contains, but in what order.

So there is a need for a formalism that shows how the atoms are linked together. This is where the sticks, representing bonds, come in. The two versions of C_2H_6O, which are called *isomers* (same components, different order), can be represented in this way:

```
    H         H             H    H
    |         |             |    |
H — C — O — C — H       H — C — C — O — H
    |         |             |    |
    H         H             H    H

  Dimethyl ether             Ethanol
```


A further complication is that the molecule occupies, not the two-dimensional plane of a page, but all three dimensions of space.

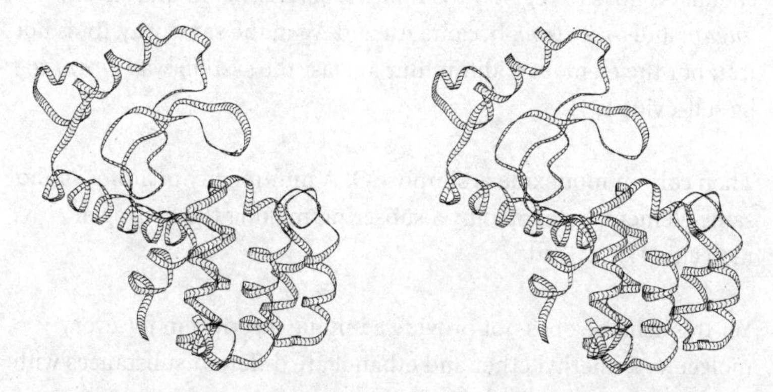

7. This pair of stereoscopic images allows molecular shapes to be seen in three dimensions. The molecule shown here is a form of the enzyme lysozyme, which is present in saliva. Coiled (helical) sections of the protein chain are clearly visible here. Place the page about 8 inches from your eyes, and cross your eyes slightly so that you can see three images. Focus attention on the middle one – in a few seconds, it should become sharp.

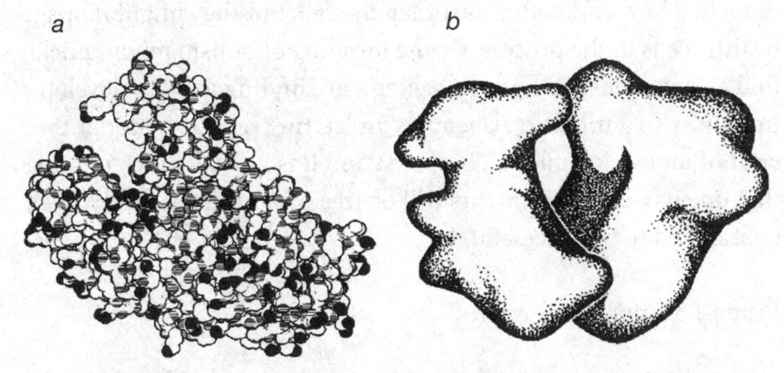

8. *a*: Space-filling representations of molecules show how they occupy space. This is the enzyme molecule DNA polymerase, which constructs new DNA molecules. The shading distinguishes different kinds of atom. *b*: Incomplete knowledge of atomic-scale structure, or simply desire to avoid too much detail, can sometimes necessitate very schematic depictions. Here I show the ribosome complex, which constructs new proteins.

There are various ways of accommodating the third dimension, which have become more sophisticated with the advent of computer graphics. Fig. 7 (see page 22) shows a stereoscopic representation of a medium-sized molecule: with practice you can get the two images to overlap and see the shape in 3D.

This by no means exhausts the schemes that chemists have devised. Sometimes there is a call for 'space-filling models' that reveal how great a volume of space the molecule occupies (Fig. 8*a*). Sometimes *ad hoc* schematizations work best, to avoid showing unnecessary detail (Fig. 8*b*).

Making molecules

The molecules in Fig. 8 are biomolecules: fantastic constructions painstakingly assembled by the cell's machinery so that every atom sits in the right place. Chemists cannot yet approach this level of artistry, which is why nature often gets the upper hand. The molecules that we make for killing pathogenic bacteria, for foiling viruses, or for destroying cancer cells often do their job rather crudely. They work, often surprisingly well, but they might damage healthy cells in the process. Or the invading organism might quickly find a way to outwit them, as bacteria are finding ways to develop immunity to antibiotics. Chemists are getting rapidly better at the craft of molecule building, however, and it is not too much to hope that one day drug treatments will be free of side effects, and guaranteed to be successful.

Primo Levi put it this way:

> when you come right down to it, we're bad riggers. We really are like elephants who have been given a closed box containing all the pieces of a watch: we are very strong and patient and we shake the box in every direction and with all our strength. Maybe we even warm it up, because heating is another form of shaking. Well, sometimes, if the watch isn't too complicated, if we keep on shaking, we succeed in getting it together . . .

This is actually something of a worst-case scenario. Levi wrote it in 1978, and things have come a long way since then. Yet, until the last few decades of the twentieth century, the approach that Levi describes, which chemists like to call 'shake and bake', was often the best they could do. Most of the effort in synthetic chemistry is devoted to making so-called organic molecules, which means that they have skeletal frameworks built largely from carbon atoms. Most of the molecules I have depicted so far are considered to be organic molecules. In dimethyl ether and ethanol, the carbon framework is rather small. In Primo Levi's molecule (page 15), it is a more complicated skeleton. And you can see that nitrogen atoms are also part of his framework, while an oxygen atom serves as a crucial bridge in dimethyl ether. Organic molecules are not necessarily based on skeletons exclusively of carbon – just predominantly so.

'Organic' might seem a strange choice of word, for nearly all of the molecules that organic chemists play with are products not of nature's organisms but of the laboratory. The term is a historical one, for organic chemistry was once indeed the study of the molecules derived from living organisms. These, it became clear, were largely carbon-based. Why carbon? Atoms of carbon are almost unique amongst the elements in their ability to link together into stable frameworks with complicated shapes: rings, long chains, branching networks.

Chemists of the nineteenth century had at best only a dim awareness of how to make new organic molecules. They could modify the molecules that nature provided, chipping fragments off the carbon backbone and replacing them with others; but altering the framework itself was more difficult. The problem was rendered still more refractory by the fact that they usually had little idea of the true architecture of the molecule they wanted to make. It is a wonder that their shake'n'bake methods got them as far as they did, providing the first synthetic plastics, dyes, and drugs.

By starting where some of these chemists started, we can begin to see what molecule building is about, and why it is so difficult – and so desirable. In the mid-1850s, the German chemist August Wilhelm Hofmann, working in London, directed his teenaged student William Perkin to make the compound quinine from the distilled components of coal tar. Quinine is a natural extract of the cinchona tree, and was used to treat malaria. Coal tar was a sticky black residue produced in great quantities by gas works, which sprang up in the early part of the century following the invention of gas lighting. It was unpromising stuff, but Hofmann and others had discovered that by distillation one could separate from it several carbon-rich, odorous ('aromatic') organic compounds such as benzene, toluene, xylene, and phenol.

No one knew the structures of any of these compounds – no one could draw stick diagrams like those shown earlier, indicating the connectivity of the atoms. All they knew was how much of each element the compounds contained, and thus what its chemical formula was. Benzene, for example, has the chemical formula C_6H_6, and quinine $C_{20}H_{24}N_2O_2$. The shape of the carbon backbones in these molecules was totally unknown.

Hofmann's procedure (and therefore Perkin's) was to count atoms. They began with a coal-tar extract that they could convert to a compound called allyltoluidine with most of the right atoms in roughly the right ratios, and hoped that some appropriate treatment of this substance would convert it to quinine. They guessed that two molecules of allyltoluidine (with formula $C_{10}H_{13}N$) might combine with some oxygen and hydrogen to make the drug. But it was a long shot, for there are many ways of linking ten carbon atoms together; and, as it happened, the carbon framework of allyltoluidine is not the same as that of half a quinine molecule.

So the experiment, which Perkin conducted in a laboratory rigged up at his parents' house in east London, did not work – it just gave a rust-coloured sludge, something unhappily familiar to organic chemists. Yet the young Perkin did not give up – he tried instead starting with an organic compound called aniline in place of allyltoluidine. This time the sludge was black. But when it was dissolved in methylated spirits, it produced a glorious purple colour, which, to Perkin's delight, would dye silk. He had discovered the first aniline dye. He set up a factory with his brother and father to make the stuff, and it was soon being manufactured in large quantities in Britain and France. This marked the beginning not only of the synthetic dye industry but of the entire modern chemicals industry – for many of today's chemicals companies, such as BASF, Ciba-Geigy, and Hoescht, began as manufacturers of aniline dyes.

By the last quarter of the century, the synthesis of organic molecules had become a less haphazard business. August Friedrich von Kekulé deduced in 1857 that carbon atoms are four-valent – they like to form four bonds. And in 1865 he proposed that benzene, to which all of the aromatic coal-tar molecules were related, contains a *ring* of six carbon atoms – an ubiquitous leitmotif of organic chemistry. In 1868 the German chemists Carl Graebe and Carl Liebermann synthesized the alizarin molecule, which is responsible for the red colour of the dye extracted from the root of the madder plant. This was one of the most commercially important of all natural dyes, and Graebe and Liebermann's synthesis eventually made it available much more cheaply as an artificial product.

The synthesis of alizarin stands as a landmark in molecule making for two reasons. First, it was achieved by planned modification of the starting material (another coal-tar aromatic compound called anthracene), rather than by cooking up the ingredients and hoping for the best. The chemists had some knowledge not only of the formula but also of the chemical structure of anthracene, which they knew to be related to that of alizarin. (In fact they guessed the wrong structure, but fortunately this turned out not to matter.) Organic chemists call this kind of procedure – in which a starting molecule is converted systematically, bit by bit, to the desired product – a *rational synthesis*.

Secondly, by making alizarin in the laboratory Graebe and Liebermann had shown that organic chemistry now had a prowess to rival nature's. It was possible to make the complicated molecules found in living organisms, which chemists today call *natural products*.

So was the synthetic red dye made by Graebe and Liebermann, and later by the ton in chemicals factories, identical to the natural red madder dye? Yes and no. The extract traditionally obtained from madder root is a mixture of several different compounds. Alizarin is the main colourant molecule, but the extract also contains a closely related compound called purpurin, which (despite the name) imparts an orange colour. The process that converted anthracene to synthetic alizarin also generated several side products, mostly molecules with structures very similar to alizarin. Perkin was one of the chemists who, in the early 1870s, identified at least four side products in the alizarin manufactured industrially. There were doubtless many others present in even smaller quantities.

So, whereas the alizarin molecules made synthetically were identical to those extracted from madder root, the actual synthetic dyestuff was different from the natural dye – and both were impure. This is true of just about any 'natural-product' compound manufactured industrially, because all the synthetic procedures used by organic chemists generate side products. It does not necessarily mean that synthetic chemicals are better or worse than the equivalents extracted from natural sources: both are likely to be impure to some degree. But chemists place great value on purity, and spend much time ridding their products of impurities. Natural extracts, in contrast, are complex mixtures unless processed to separate the components.

The commercial value of substances derived from coal-tar compounds was not restricted to dyes. Paul Ehrlich, a German medical scientist, used the new synthetic dyes in the 1870s for staining cells, making them easier to study under the microscope. He noticed that some dyes would kill the bacterial cells that absorbed them, which suggested therapeutic possibilities. He began to synthesize dye compounds to test as drugs, and in this way he found in 1909 an arsenic-containing dye that would kill the parasite that causes syphilis. As Salvarsan, this drug offered the first relief from the deadly disease since the medieval use of mercury. It was the beginning of modern chemotherapy.

Nineteen years later Alexander Fleming discovered penicillin, a compound produced by a mould, which killed bacteria. This was the first antibiotic, and it revolutionized surgical medicine by greatly reducing the risk of wound infection. Many other natural products have beneficial physiological effects: salicylic acid, for example, an extract of willow bark, has both antiseptic and analgesic (pain-killing) properties, and a closely related molecule provides the drug aspirin, manufactured by Bayer since 1899. Chemists and medical scientists continue to comb through nature's arsenal of molecules for potential drugs, and then to find ways of synthesizing those that work.

One of these that has gained fame in recent years is the compound called paclitaxel, better known by the trade name of taxol. It is a natural product of the Pacific yew tree, and was found in the 1980s to be highly effective at preventing cells from dividing. This makes it a potential anti-cancer agent, since cancer is the result of uncontrolled proliferation of cells. It has been approved by the US Food and Drug Administration for the treatment of breast, lung, ovarian, and prostate cancer. But the Pacific yew does not provide a reliable source, since each tree yields up only a few milligrams of the compound, and it has to be extracted from the bark, so that the tree dies. Already an endangered species, it would be wiped out before meeting the global demand for taxol. There is clearly a call for synthetic taxol.

The molecular structure of taxol is fiendishly complicated. Its backbone consists of four rings of carbon: one with four atoms, two with six, and one with eight (Fig. 9). Various subsidiary groups of atoms dangle from this framework. There is no standard chemical reagent available that has this skeletal form – it must be constructed from scratch.

This is a profound challenge for the chemical rigger. Primo Levi's character gives an indication of the way that a synthetic organic chemist today would go about it:

> as you can imagine, it's more reasonable to proceed a bit at a time, first attaching two pieces, then adding a third, and so on. It takes more patience [than shake'n'bake], but actually you do get there first. And most of the time that's the way we do it.

A synthesis like this has to be carefully planned. The most common

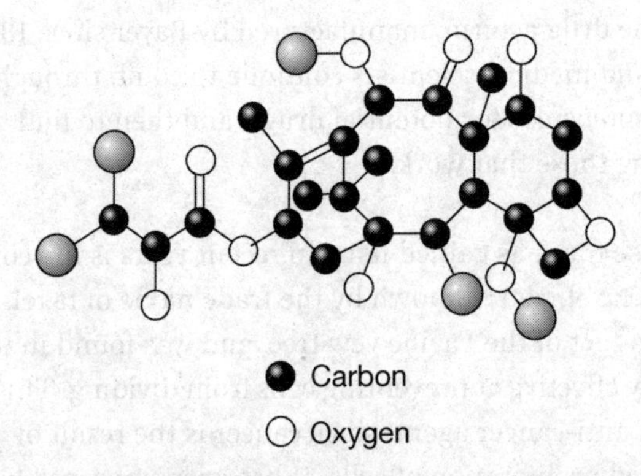

9. The taxol molecule. Here I show only the key elements of the framework: the black spheres denote carbon atoms, and the white spheres oxygen. The larger grey spheres are 'substituent' groups containing carbon, oxygen and hydrogen, which I have declined to show in detail. There are other hydrogen atoms in the molecule too, which I have omitted for the sake of clarity.

method of planning is that devised by the American Nobel laureate Elias J. Corey, who called it 'retrosynthetic analysis'. As the name implies, you work conceptually backwards from the finished product, as if disassembling a model of the molecule. At each step you break bonds that you can see how to forge, so that, when it comes to conducting the forward process, you have already seen how each link is to be made. The trick is to work back to starting materials – fragments of the carbon framework – that are readily available, or that can easily be synthesized from off-the-shelf compounds.

In the case of taxol, two groups 'got there first'. A team at the Scripps Research Institute in California, led by K. C. Nicolaou, and a group working under Robert Holton at Florida State University, described multi-step synthetic procedures within a week of one another in 1994. There is no unique way to make a molecule this complex, nor indeed a 'best' way – several alternative pathways have since been reported. But they all remain too complex to be viable for large-scale production, and taxol is currently manufactured 'semi-synthetically': from an intermediate compound found in the

10. A molecular 'big wheel' made by Achim Müller's group in Germany (top and side views). Each pyramid is a cluster of molybdenum and oxygen atoms.

yew needles, which is like half-built taxol. The synthesis can be completed relatively efficiently in the laboratory, and the needles can be removed without killing the trees.

I have talked here about making organic molecules, but I should emphasize that many chemists build molecules based on elements other than carbon. These are often rather small molecules, since other elements do not so readily form the large, intricate frameworks that carbon will adopt. One of the more remarkable exceptions to this rule is the molecule shown in Fig. 10, a ring of mostly atoms of molybdenum and oxygen. It was made by Achim Müller's group at the University of Bielefeld in Germany, and measures fully four nanometres across (that is, about fifteen times the width of a water molecule, and tens of thousand times smaller than the width of a human hair). When metals and oxygen combine, they do not usually stop in the middle ground of large molecules: either they will form molecules of a few atoms each or they will crystallize into mineral-like solids ('infinite molecules', if you like). Chemists have recently become very interested in large *inorganic* molecules like the ones shown here, because they can show unusual and possibly useful behaviour such as magnetism or electrical conduction. Components such as transistors on microchips are made of inorganic materials, primarily silicon and silicon dioxide. Tailor-made components for a molecular-scale electronics technology are just one of the items on the menu of today's sophisticated molecular cookery.

Chapter 2

Vital signs: the molecules of life

It is a little comforting when scientists and poets reach the same conclusions. Pondering the question 'What is life?' in 1949, the British biologist J. B. S. Haldane began by confessing:

> I am not going to answer this question. In fact, I doubt if it will ever be possible to give a full answer, because we know what it feels like to be alive, just as we know what redness, or pain, or effort are. So we cannot describe them in terms of anything else.

Emily Dickinson was more concise:

> Nature is what we know
> Yet have no art to say.

Yet Haldane was prepared to venture further:

> life is a pattern of chemical processes. This pattern has special properties. It begets a similar pattern, as a flame does, but it regulates itself as a flame does not . . . So when we have said that life is a pattern of chemical processes, we have said something true and important . . . But to suppose that one can describe life fully on these lines is to attempt to reduce it to mechanism, which I believe to be impossible.

Is life mere molecules, acting together with awesome but in principle explicable complexity? Or is something more involved? We simply do not know. Scientists take a bottom-up approach, assuming the minimum and invoking only testable hypotheses. Whether this will eventually lead to a point beyond which science is powerless to proceed, we cannot yet say. But no such point has so far become apparent. It seems possible that life – which we might loosely define as an organism that can reproduce, and respond to and extract sustenance from its environment – may be nothing but molecules and their relationships. Indeed, this seems extremely likely. It need not be disappointing; quite the contrary, it would be remarkable. That a conspiracy of molecules might have created *King Lear* is a possibility that makes the world seem an enchanted place.

I do not think it likely, however, that the human mind (let alone the wonders it concocts) will ever be *explained* in molecular terms, any more than *Lear* is explained by the alphabet. Most scientists do not believe so either. Phenomena are hierarchical: all things cannot be understood by considering only what transpires on a single rung. No matter how well I understand the way a transistor works, I will not be able to deduce from this knowledge why my computer crashes. If I sow seeds that fail to grow, I will do better to begin by thinking about the nutrient content, humidity, and temperature of my soil than by performing a genetic analysis of the seeds. Much of the skill in doing science resides in knowing where in the hierarchy you are looking – and, as a consequence, what is relevant and what is not.

It is worth spelling this out before we explore the molecules of the living world, because a molecular view of biology is often branded as reductionistic – as an attempt to explain every aspect of life at the molecular level of genes. This is indeed sometimes the best way to proceed, for molecules are after all the smallest functional units on which life is founded. But if we accept, as most scientists do, that by descending the ladder to the microworld we must forgo a whole range of questions and answers about life (such as: what is consciousness?), then there seems to be nothing obviously objectionable about the descent.

Indeed, this path has led us to a clearer understanding of our fundamental nature. Molecular biology has helped to fill the major gap in Charles Darwin's evolutionary theory: the issue of the *mechanism* of natural selection. It has given us at least some inkling of how life came into being on a planet of gas, rock, and water. It has saved lives and relieved much pain and suffering. It has helped us to understand why medicines do not always work as we might hope, why irresponsible use of antibiotics has bred superbugs, how the AIDS virus does its terrible work. The study of life's molecules became the major science of the twentieth century's second half, and looks set to have an ever greater impact on our lives in the future. It is perhaps the one area of science in which some degree of knowledge is no longer a luxury.

The vital force

Organic chemistry was once considered different from the rest of chemistry. Many scientists believed in the early nineteenth century that organic matter was the product of a vital force operating in living organisms, which the chemist could never mimic in the laboratory. But by 1818 the influential Swedish chemist Jons Jacob Berzelius saw tautology looming in the idea of vitalism, while despairing of ever getting beyond it:

> the cause of most phenomena within the Animal Body lies so deeply hidden from our view, that it certainly will never be found. We call this hidden cause *vital power*; and like many others, who before us have in vain directed their deluded attention to this point, we make of us a *word* to which we can affix no idea.

Yet in the same breath, as it were, Berzelius hinted at how to go further:

> This *power to live* belongs not to the constituent parts of our bodies, nor does it belong in them as an instrument, neither is it a simple power; but the result of the mutual operation of the instruments and rudiments on one another . . .

Here is the key. Understanding the molecular basis of life is not so much about appreciating what the molecules are, as what they do to one another. The molecular nature of life is not a gallery but a dance. In later chapters I shall describe some of the steps; here I want briefly to introduce some of characters.

In Haldane's time it was not unusual to regard life as a series of chemical transformations being conducted as if in some vast network of laboratory glassware. The key to it all, scientists believed, was metabolism: how we obtain energy from food. But you will not make an organism by throwing into a pot all of the purified molecular components of the cell. A modern view of molecular biology is concerned with *organization* in time and space. How do the molecules of life arrange themselves amongst the cell's compartments, how are they shifted around, how do they communicate so as to synchronize their action? We can ask these questions only because we can now inspect the working cell at the molecular level, taking measurements and snapshots of molecules going about their business. And so the cell becomes a community.

Yet it is a community of Byzantine complexity. Molecular biology is not difficult in the way that theoretical physics is difficult – the concepts are not unfamiliar, abstract, or mathematically abstruse. The difficulty arises because there is so much going on all at once. We react with surprise and shock when things go wrong with our own molecular machinery, but it is far more astonishing that the machinery works at all. Frequently it does so because it is designed to be robust in the face of the world's vicissitudes. There are checkpoints, safety mechanisms, back-up plans, careful record-keeping. No human-made mechanism is anything like as sophisticated or as well organized as a cell.

It is important to bear in mind that the cell is a community of automata. Its members have no volition, no foresight, no memory, no altruism (nor selfishness, in the strict sense). They often collaborate so beautifully that it is easy to forget this. On the other hand, cells can be unpredictable, because we know so little about how they work. They might survive when we expect them to die, or they might react to a potential drug in totally unforeseen ways.

Molecular biology works at the level of the cell, and seldom talks about the whole organism. The cell is the 'atom of life' – you cannot get any smaller and still be alive. (Viruses are a debatable exception – they are little more than genes wearing a coat, but they cannot reproduce without hijacking the machinery of the cells they infect.) This need not be as restrictive a viewpoint as it might seem, since we can understand an awful lot of our requirements as humans according to what goes on in a single cell. A human cell needs oxygen and sugar to make new molecules and to replicate itself – so we breathe and eat. Nerve impulses start at the level of the cell. Our tissues – skin, hair, bone, muscle – are put together molecule by molecule in the cell. We excrete to remove the cell's waste. We shiver and sweat to stabilize our cells' temperature. Very many questions about the way we function can, in other words, be addressed on the rung of molecular biology. Of course, amongst those that cannot are many of the most interesting.

The players

Haldane attributes to Engels the proposition that life is 'the mode of existence of proteins'. (He was a socialist, and did not generally read Engels for the biology.) This vitalistic statement implies that proteins are inherently alive, an idea that Haldane squashes. But he has no objection to the idea that *proteins* are the stuff of life.

Proteins are substances found ubiquitously in living cells. Many of them are enzymes, molecules that catalyse processes of chemical change. Enzymes speed up chemical reactions by factors of perhaps several millionfold and thereby ensure that the body's chemistry is not impossibly slow. They were discovered from studies of fermentation: *enzyme* is Greek for 'in yeast'. In the late nineteenth century it was found that enzymes could be extracted from yeast cells and purified, yet would remain capable of bringing about fermentation despite the fact that they were no longer part of a living system. This discovery helped to establish that the chemistry of life works according to the same principles as the chemistry of non-living matter.

If the cell is a city, enzymes are the workers. To keep the city running, raw materials are imported and converted into useful items. Enzymes populate the cellular factories in which this is done. One curious aspect of this manufacturing industry is that it includes factories for making the workers themselves: enzymes too are put together on a production line.

Not all proteins are enzymes. Some serve a structural role, providing the tissues of the body. Some act as the cell's police force, others carry packages to and fro in a protein shuttle service running on protein tracks. Some operate the portals to the cell, sitting in the outer membrane and opening or closing in obedient response to the instructions they receive. There are something like 60,000 different varieties of protein molecule in human cells, each conducting a highly specialized task.

It would generally be impossible to guess what this task is merely by looking at a protein. They are undistinguished in appearance, mostly globular in shape (see Fig. 8, page 23) and composed primarily of carbon, hydrogen, nitrogen, oxygen, and a little sulphur. All the proteins that fulfil a particular task have the same shape and structure – the seemingly amorphous blob is in fact exquisitely designed and assembled.

Many enzymes are shaped a little like a knobbly kidney bean, with a cleft in the inner curve. This cleft is where the action is – where the molecule performs its catalysis. Some proteins do their jobs in groups: they become 'subunits' of a many-protein assembly. The enzyme tryptophan synthase, which bacteria possess, is one of these, built from four detachable subunits. This enzyme synthesizes the small molecule tryptophan, which is essential to all organisms. Humans do not possess the enzyme, and so we have to get our tryptophan ready-made by eating organisms that have constructed it.

As this example implies, enzymes and other proteins are commonly given names that reveal their function. Alcohol dehydrogenase is an enzyme that takes a hydrogen atom from ('dehydrogenates') an alcohol molecule. ATP synthase synthesizes the molecule ATP. But not all protein names are so transparent. Haemoglobin, which carries oxygen in the bloodstream, gets its name from the Greek for blood (*haeme*) along with the fact that it is globular. Myoglobin, to which haemoglobin donates its oxygen cargo in muscle tissue, derives the first part of its name from the Greek word for muscle. Other names are more whimsical. Elastin is an elastic protein found in many flexible body tissues, such as blood vessels and vocal cords. Ubiquitin is a protein found just about everywhere in the body, because it plays a central role in the universally necessary process of destroying obsolete proteins.

You would not guess, from looking at pictures like Fig. 8, that a protein molecule is in fact a single chain of small molecules linked together. The chain is folded and coiled on itself so densely that it looks like just a mass of atoms. But close inspection of the structure obtained from X-ray crystallography (see page 18) allows us to follow the strand as it twists and turns through the compact globule. Protein chemists sometimes show this strand explicitly in a different kind of representation of a protein (Fig. 11). Here you can see that the structure is built up from certain repeating features or 'motifs', such as coils, which are called alpha helices, and so-called

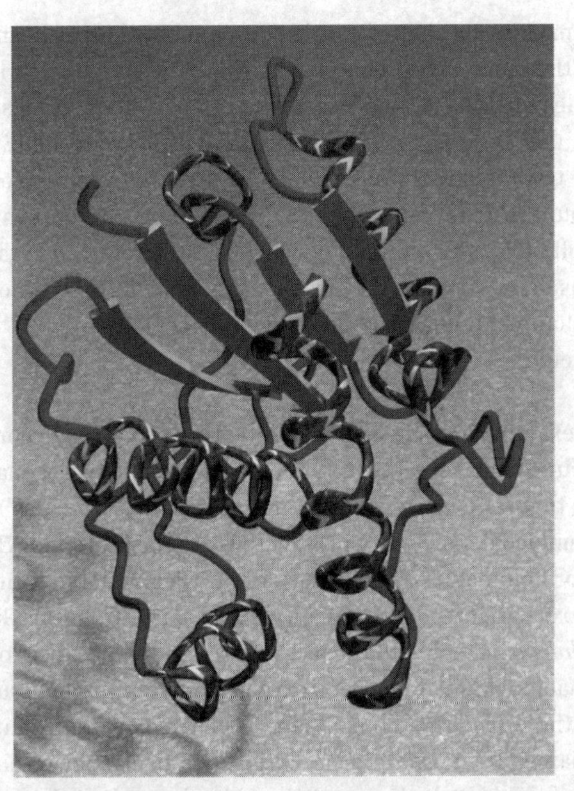

11. In this representation of a protein molecule, its folded-chain structure is made explicit. The flat, roughly parallel arrows denote beta-sheet motifs, while the coiled segments bearing chevrons are alpha helices. The molecule is the enzyme phosphotyrosine protein phosphatase from a cow.

beta sheets, where several parts of the strand lie parallel to one another.

A protein's structure can be conceptually decomposed even further. The chain is made up of small characteristic clusters of atoms, joined in a sequence like beads on a string. These clusters were once separate molecules, called *amino acids*. There are twenty varieties of amino acids in natural proteins. In the chain, one amino acid is linked to the next via a covalent bond called a *peptide bond*. Both molecules shed a few extraneous atoms to make this linkage, and the remainder – another link in the chain – is called a *residue.* The chain itself is termed a *polypeptide.*

Any string of amino acid residues is a polypeptide. We can make them ourselves simply by heating up a mixture of amino acids. But we will not make a protein this way. In a protein the order of amino acids along the chain – the *sequence* – is not arbitrary. It is selected (that is, naturally selected, in Darwin's sense) to ensure that the chain will collapse and curl up in water into the precisely determined globular form of the protein, with all parts of the chain in the right place. This shape can be destroyed by warming the protein, a process called denaturation. But many proteins will fold up again spontaneously into the same globular structure when cooled. In other words, the chain has a kind of memory of its folded shape.

The details of this *folding process* are still not fully understood – it is, in fact, one of the central unsolved puzzles of molecular biology. We do know, however, what holds the polypeptide chain in its compact form in a protein molecule. Many parts of the chain are able to form weak bonds with each other, called *hydrogen bonds.* These glue the chain into alpha helices and beta sheets, for example. And some parts of the chain are held together by stronger bonds forged between sulphur atoms, which dangle from the residues of the amino acid cysteine. Some residues are relatively insoluble in water, and these will tend to gather together in the core of the protein globule, surrounded by more water-soluble parts of the chain. So the folded structure that results depends on the character of the various residues, and where they are located along the chain – in other words, on the sequence. You could say that protein molecules are manufactured with their own folding instructions.

How does the cell's machinery 'know', when making a protein, in which order the amino acids should be strung together? This is where DNA (deoxyribonucleic acid) enters the picture. The sequence of every protein in the body is encoded in the DNA molecules that reside in every cell. Proteins do all the work (or most of it); DNA sits passively waiting to be read, when the need for a protein arises.

DNA information is written in a different language from protein information, but the cell can translate between the two. DNA is another string of small molecular units – another *polymer*. But its building blocks are different. Rather than amino acids, they are molecules called *nucleotides*. DNA is a library of protein structures, written in the characters of the *nucleotide sequence* (see page 135). The information for making each protein is, roughly speaking, encoded in a sketch of DNA called a *gene*.

There is clearly some impressive coordination involved here. Whenever a job needs doing by enzymes, the message must get through to the region where the DNA resides. In human cells and those of all other organisms other than the most 'primitive' single-celled bacteria, this is a central compartment called the *nucleus*, which is fenced off within its own membrane (Fig. 12). Organisms with cell nuclei are called *eukaryotes*.

In human cells DNA is packaged up in bundles called *chromosomes*. To make a protein, the stretch of DNA containing the corresponding gene is unravelled and read. In fact, proteins are made not in the nucleus but in a different compartment called the endoplasmic reticulum, a labyrinthine network of membrane channels. The gene is *transcribed* first into a molecule related to DNA, called RNA (ribonucleic acid). The RNA molecules travel from the nucleus to the endoplasmic reticulum, where they are *translated* to proteins. The proteins are then shipped off to where they are needed. So the cell's molecules must be capable of communication and transport.

The whole affair is regulated so that proteins are not being made willy-nilly, but only on demand. If the cell was constantly making all the proteins at its disposal, it would rapidly seize up. A key insight

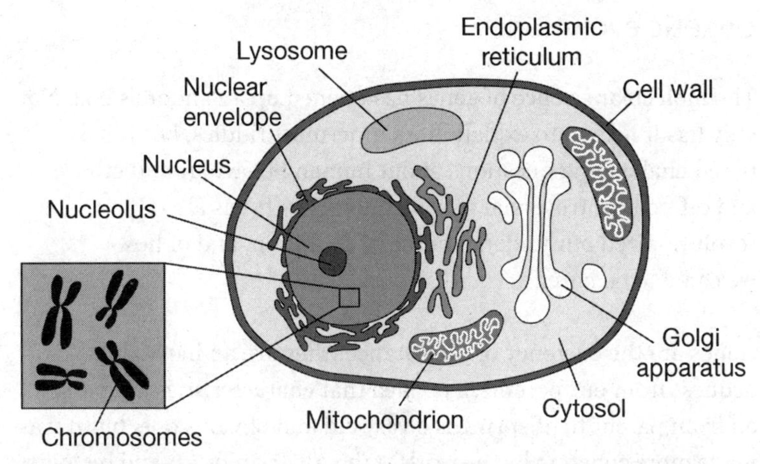

12. Cells of humans and other eukaryotic organisms sequester their genetic material (DNA) in a central nucleus. Various other compartments (organelles) perform a variety of other functions, such as protein synthesis and energy production. Note that the DNA is bundled into chromosomes (as shown here) only when the cell is about to divide; at other times it is unravelled into thin strands.

into how the cell maintains order amongst its component parts emerged from the work of French biochemists François Jacob and Jacques Monod in the 1960s. They showed that genes regulate one another, switching each other on and off via the agency of the proteins they encode. For example, some genes that encode proteins used in the cell (called *structural genes*) are partnered by *regulatory genes* that encode repressor proteins. When the regulatory gene is switched on, the repressor protein is synthesized and binds to the structural gene, preventing it from being 'expressed' (transcribed and translated into its protein). Jacob and Monod called these regulated stretches of DNA *operons*. They provide just one illustration that the cell employs a weblike network of interactions between different genes and proteins. Molecular biologists have now decoded more or less all of the nucleotide sequence of human DNA – but they have so far mapped out only very small regions of the web that it weaves.

Genetic evolution

The molecular science of genes has opened up a Pandora's box. Not only has it helped to explain life's innermost riddles, but it has posed challenging questions about human behaviour and ethics, and offered controversial new technologies. It has also revolutionized our understanding of evolution, and of how we came to be here.

Genes are the currency of inheritance: they are an inevitable bequest from our parents. The idea that characteristics are passed on from parent to offspring is a very old and obvious one, but it was made more concrete by the work of the Austrian priest and biologist Gregor Johann Mendel in the nineteenth century. His experiments on the propagation of peas led him to suppose that there are 'particulate factors' that mediate heredity, passing from the cells of the progenitors to their progeny. It soon became clear that these 'factors', later called genes, were molecular in nature, but throughout the first half of the twentieth century many scientists thought they were protein molecules. Haldane, as we have seen, shared this belief. Not until Francis Crick and James Watson deduced the structure of DNA in 1953 was there a compelling case for regarding DNA, rather than proteins, as the molecular stuff of heredity, the fabric of genes.

This placed evolution on a concrete molecular basis – for what a fertilized egg gets from its parents is not a preformed body but a set of genetic instructions for a body plan. The evolutionary changes that happen slowly from generation to generation are due to changes in the molecular make-up of the genes. DNA is copied when a cell divides – but not always perfectly. So the DNA that a child gets from mother and father may be a slightly flawed amalgam of both their genes. Generally these flaws will not matter. Sometimes they will be harmful (but note that most genetically based diseases are the result of a child *inheriting* a faulty gene, not acquiring one from random copying errors). Very rarely, a genetic mutation will have a beneficial effect, making the organism better equipped for survival. The advantage might be extremely slight – but evolution advances through such infinitesimal steps, as tiny advantages lead to fractionally higher reproductive success and thus to a slow increase in the incidence of the mutated gene in the population.

What this all means is that genes are a molecular record of evolution. The common ancestor of humans and rabbits shared the same set of genes. The differences that now exist between the respective total complement of genes (the *genomes*) of humans and rabbits reflect the divergence attributable to an accumulation of genetic mutations. This enables scientists to reconstruct evolutionary histories – to deduce the order in which species diverged – from the molecular structure of genes. Previously, palaeontologists had to make such deductions on the basis of the shapes of bodies or bones; now they have a more readily quantifiable molecular measure of evolutionary change.

In particular, evolutionary trees or *phylogenies* can be reconstructed by comparing the special DNA that is housed in a compartment of cells called the mitochondrion (see Fig. 12). This is the cell's furnace, where the energy is produced (see Chapter 4). Unlike the DNA in the nucleus, mitochondrial DNA comes directly from the mother. Whereas nuclear DNA is changed by reshuffling the genes of both parents, mitochondrial DNA is changed only by the gradual accumulation of mutations from generation to generation – so it is a better record of evolutionary change. In 1987 Allan Wilson of the University of California at Berkeley and co-workers compared the mitochondrial DNA of people from many racial groups. Knowing the average mutation rate, they figured out that all of the samples, and by extension the mitochondrial DNA in all living humans, derived from a single version that existed 200,000 years ago in the cells of an African woman – the common ancestor of all of humanity. Molecules contain a record of history richer than anything to be found in fragments of clay pots or ancient burial mounds.

The RNA world

All living organisms, from the humblest of bacteria to the most regal of kings and queens, have their genetic material packaged in DNA, and put into effect by proteins. This implies that all life has a common origin.* The most primitive single-celled organisms must have contained proteins and DNA very similar to those in 'simple' bacteria today.

But what came before? The molecular symbiosis whereby DNA encodes proteins and proteins help DNA to function and to replicate is fantastically sophisticated even in a bacterium. Neither proteins nor DNA could have come spontaneously into existence from fragments of organic molecules scattered throughout the seas and lagoons of the early Earth: their structures are just too complex to have assembled at random. It is easier to understand (at least in principle) the 3.8 billion years of evolution from the earliest bacteria or algae to the present day than to understand how, over maybe just a few hundred thousand years, Earth changed from a barren planet to one that cradled life.

Chemists have devised many inventive schemes according to which the inorganic constituents of the young Earth, such as methane, carbon dioxide, ammonia, water, and nitrogen, might have become transformed into the amino acids and sugars needed to make the molecules of life. They are all tentative; no theory yet prevails for the chemical origin of life. But the conceptual difficulties in progressing from rock, gas, and water to prototypes of biomolecules

* We simply do not know whether the DNA–protein partnership is the *sine qua non* of life, and it would be rash to suggest that life can have no other molecular basis. But none has been found. It is not so difficult to imagine that slight, systematic modifications to DNA could give rise to an alternative genetic system – but no organisms show such a thing.

are still smaller than those of turning these building blocks into functioning cells full of proteins and DNA. It is a chicken-and-egg problem: on their own, both proteins and DNA are useless.

The favourite way out of the conundrum is to switch attention to the humble go-between: RNA, which carries the genetic information to the machinery of protein synthesis. RNA is much more versatile than DNA. In the 1980s, RNA molecules in the cell were discovered that could act as catalysts for their own rearrangement. Human genes are corrupted with a lot of 'junk' that needs to be excised before the message can be clearly read (see page 140). This junk gets copied into RNA, but is then snipped out before the RNA is translated into a protein. This editing is largely conducted by enzymes; but some RNA molecules can do it unassisted. These are called *ribozymes*, reflecting the fact that they show enzyme-like tendencies.

During the 1990s, biochemists greatly expanded their appreciation of RNA's abilities. Using the techniques of biotechnology developed for manipulating and rewriting DNA, they have made synthetic RNA molecules that can conduct all manner of chemical processes, such as linking together nucleotides or forming bonds between carbon atoms. These studies show that in principle RNA is versatile enough to bring about many of the chemical transformations that would have been necessary for life to begin. In short, RNA can act both as a gene-carrier and as a worker.

Many scientists researching into the origin of life therefore postulate an era that they call the *RNA world*, which existed before the double act of proteins and DNA appeared. RNA is still a fiendishly difficult molecule to make under conditions comparable to those on the early Earth. But the RNA world breaks the impasse posed by the mutual dependence of proteins and DNA, and so provides a conceptual link between the formation of small organic molecules and the appearance of the first primitive cells.

Synthetic life

If we do eventually come to understand the origin of life – not necessarily how it really began, but at least how it *might* have begun – could we then rerun it in the laboratory? Can we create life from scratch?

Enough is known already about the molecular basis of life for researchers to be able to speculate about building an artificial cell. It sounds perhaps like a frightening prospect. What if we happened to make a cell that was far better at replicating than 'natural' cells? Would they colonize the planet like an alien invasion?

This is not science fiction. In fact, I think that a synthetic cell will be made within the twenty-first century, for better or worse. It is already routine to make synthetic DNA and to rewrite genes, and many chemists are working on building 'designer proteins' from scratch. The biochemists Jack Szostak, David Bartel, and Pier Luigi Luisi have proposed that:

> Advances in directed evolution and membrane biophysics make the synthesis of simple living cells, if not yet foreseeable reality, an imaginable goal.

They suggest that a 'minimal cell' could be constructed from tailor-made ribozymes. A primitive version of a ribozyme that can assemble RNA (and so potentially replicate itself) has already been reported. These could be encapsulated within artificial membranes like those of cells, but with an ability to grow and divide: Luisi has made such 'replicating membranes'. The replicating 'protocells' might evolve RNA molecules capable of assembling amino acids into proteins. This would then allow us, say the researchers, to 'replay the tape of early evolution'.

Those who foresee terrible purposes in such experiments might bear in mind that this is an absurdly difficult way to try to develop some deadly weapon, when lethal chemical and biological weapons can be made already with relative ease. All the same, it is impossible to know where such research might lead. That is the reality of molecular science: it is a creative discipline, which gives us something to show for our efforts at the end. Therein lies all the artistry, all the wonder, and all the peril. In the end, we will only get the molecules we deserve.

Chapter 3
Take the strain: materials from molecules

The hardest part of space travel (apart from the boredom and the danger) is the leaving. In the vacuum of space, free from strong gravitational influences, a small burst of propulsion will keep a rocket moving almost indefinitely. So most of the fuel that a rocket takes on board is needed simply to escape the Earth's gravity. This fuel and the engines that burn it account for the greater part of a rocket's mass. I recall vividly the Apollo missions that left the Earth as a gleaming tower and returned as a tiny nub from the nose.

If we could launch spacecraft from outside Earth's atmosphere, the payload would therefore be much diminished. In his novel *The Fountains of Paradise*, Arthur C. Clarke suggested how this might be done. He posited the Space Elevator: a platform positioned in geostationary orbit around the Earth, tethered to the ground by a long, superstrong cable. Space hardware and any passengers are shuttled up by elevator to the platform, from where they can be launched into space with a fraction of the fuel requirements needed for a ground-based takeoff.

To tether an orbiting platform to the Earth's surface, we would need cables far stronger than anything currently available. They would need to be lightweight too – the weight of that much steel cable would be immense.

It does not take the speculative concept of a Space Elevator to explain why we need materials that are strong, tough, corrosion-resistant, lightweight, and so forth. But scenarios like this serve to motivate the question of just how far one can go in improving materials' properties. Superstrong 'space tethers' are also being considered for launching objects into space by a kind of slingshot process in which the payload is temporarily attached by a thread to an orbiting satellite. Such tethers have to be both lightweight and strong. And there is no lack of more mundane demands for strong cables: for example, to hang bridges and tether drilling rigs to the seabed.

Tough fibres have always been available to us, for we are fortunate that nature supplies them in abundance: silk, hemp, wood, hair. In pre-revolutionary Russia silk was used for bulletproof protection, and artificial silk is being developed today with the same purpose in mind.

The plastic age, which began in earnest in the early twentieth century, has supplemented natural fibres with synthetic ones that have both advantages and shortcomings. The earliest plastics were made by trial and error; modern plastics, in contrast, are designed at the molecular level for the applications they will serve. In this chapter on the molecular aspects of materials, I will focus my discussion on fibres both natural and human-made, since these provide some particularly beautiful and graphic examples of how molecular structures affect the kind of material properties that engineers worry about.

But the interaction of molecular science with materials engineering is far, far broader than this. Molecules are designed that can be converted into ultrahard ceramic materials capable of withstanding high temperatures. These are used in aerospace engineering, turbine manufacture, and power generation. Materials made from molecules (particularly polymers) are designed to conduct electricity, to capture, guide, and transform light pulses, to sieve other molecules, to protect surfaces from corrosion or contamination. They include 'smart' materials that respond to changes in their environment, which can act as autonomous switches, valves, and pumps. The impact of molecule-based materials on medicine is huge, and destined to be ever-more important: they provide artificial limbs, organs and tissues, systems for administering drugs, biodegradable sutures for surgery, sensors for monitoring the body's state of health. One day molecular engineering will make it possible to grow a new kidney or a new heart to replace a damaged one.

Cables of the body

The idea that our bodies are a mass of cellular communities of proteins does not tally with our experience. We experience ourselves as a composite of fabrics: skin, bone, muscle, hair, fingernails. This material framework has the properties necessary to let us interact with the world. The outer layer of skin is just cladding, made up of cells that are programmed to die once they have formed the tissue. The same is true of hair, which provides insulation, and of fingernails, the evolutionary remnants of claws that once punctured and ripped. Bone and tooth are composed largely of a hard, inorganic material: calcium phosphate. Most of these materials are continuously renewed while we live; some, such as the proteins in the eye's lens, are not.

These natural materials of the body serve mechanical roles and maintain our structural integrity. They are like the bricks, girders, and cladding of a building, which protect the workers from the elements and house all the complex wiring and plumbing that is necessary to conduct business as usual. Many of the body's structural fabrics are proteins. Unlike enzymes, structural proteins do not have to conduct any delicate chemistry, but must simply be (for instance) tough, or flexible, or waterproof. In principle many other materials besides proteins would suffice; and indeed, plants use cellulose (a sugar-based polymer) to make their tissues. Yet the marvel of proteins is that they are so versatile. The molecular chains can be woven into strong fibres; cross-linked or entangled, they form the stiff matrix of horn and claw, or elastic sheets. What is more, the raw materials for making proteins are abundant in the cell. And because proteins are encoded in genes, the molecular features that give a structural protein its mechanical properties can be delicately tuned and then reliably reproduced.

The most abundant structural protein in the human body, comprising about a quarter of our total protein mass, is collagen. This is a relatively simple protein whose chainlike molecules contain mostly two sorts of amino acids: glycine and proline. Glycine constitutes every third link of the chain, with proline and other amino acids (particularly lysine) in between. Some of the proline units are chemically modified, having an oxygen atom added. This takes place in a reaction that involves vitamin C, which is why this compound is needed to maintain healthy tissues. Lack of vitamin C leads to the condition known as scurvy, caused by damaged collagen that has not been replaced.

Collagen exemplifies the way in which natural protein-based structural materials differ from most synthetic polymer-based plastics. Both are composed of chain molecules; but in structural proteins these chains gather together in complex arrangements, forming thicker fibrils like ropes woven from string woven from thread. This kind of arrangement, in which structural elements are built up at successively larger scales from smaller components, is said to be *hierarchical*. Engineers have learnt to use the same principle: Gustav Eiffel's iconic tower, for instance, contains struts made of a network of smaller girders, some of which are built up from even smaller girders (Fig. 13).

Collagen displays several different kinds of large-scale structure,

13. Structural hierarchy, common in nature's materials, is exemplified in the Eiffel Tower.

like several different designs of tower, but all are constructed from the same basic small-scale elements (Fig. 14). Each collagen molecular chain crimples up into a helix. Three of these twist around one another to form a ropelike triple-helical 'microfibril'. These microfibrils aggregate together in various ways. For example, they can gather in a staggered arrangement to form thick strands called banded fibrils. The staggering creates the appearance of dark bands under the electron microscope. Banded fibrils constitute the connective tissues between cells – they are the cables that hold our flesh together. Bone consists of collagen banded fibrils sprinkled with tiny crystals of the mineral hydroxyapatite, which is basically calcium phosphate. Because of the high protein content of bone, it is flexible and resilient as well as hard.

Collagen fibrils are, however, not outstandingly strong on their own, because the molecules are not linked or entangled together. But other types of collagen contain cross-linked bundles of microfibrils, joined into a kind of tough web or mesh. This provides the

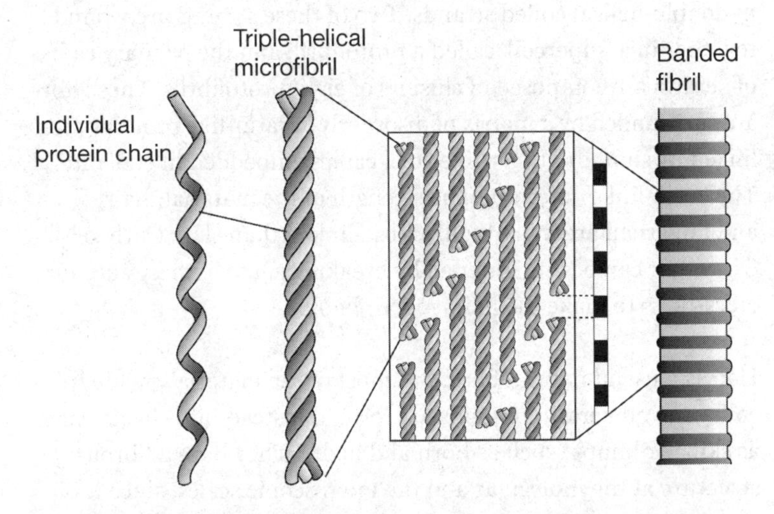

14. Collagen has a hierarchical structure of coiled coils. A staggered arrangement of these collagen microfibrils causes the dark bands seen microscopically in banded fibrils, where a metal staining agent attaches itself to the ends of the microfibrils.

membrane separating the outer layers of skin from thc inner layers.

In contrast to the disorderly tangle of connective tissue, the eye's cornea contains collagen fibrils packed side by side in an orderly manner. These fibrils are too small to scatter light, and so the material is virtually transparent. The basic design principle – one that recurs often in nature – is that, by tinkering with the chemical composition and, most importantly, the hierarchical arrangement of the same basic molecules, it is possible to extract several different kinds of material properties.

Collagen also constitutes the tough, elastic fabric of tendons and ligaments, and the reinforcing dentine of tooth. But the proteins on our outside – in skin, hair, and nail, as well as animal horn and hoof – are of a different kind. These tissues consist mostly of keratin, another hierarchically structured fibre. The molecular chains of keratin are again wound into helices, which are paired up in double-helical coiled strands. Two of these strands are wound together in a 'supercoil' called a protofibril, and the primary cables of keratin are composed of clusters of eight protofibrils. These fibres are surrounded by a matrix of disorderly keratin-like proteins cross-linked by sulphur atoms, like steel cables embedded in concrete. The cross-links determine the strength of the material: hair and fingernail are more highly cross-linked than skin. Curly or frizzy hair can be straightened by breaking some of these sulphur cross-links to make the hairs more pliable.

Hair is a useful natural fibre; but most of the materials made from collagen and keratin proteins are formed instead into sheets (such as skin) or lumps (such as horn and hoof). They have a fibrous structure at the molecular and the microscopic scales, since it is easier for the production machinery of individual cells to manufacture such structures than, say, to cast solid blocks. But the body then organizes these micro-fibres into other shapes.

Web dreams

Silk, on the other hand, shows just how impressively evolution can rise to the challenge of making fibres when they are truly called for. Here is a substance that can be spun into threads all but invisible to the passing fly, yet flexible enough to absorb the energy when the fly hits the web, and strong enough not to break under the impact (Fig. 15). Stronger than steel or the best human-made fibres, silk is admired by the engineer for its robustness and prized by the textiles manufacturer for its exotic shimmer, its cool texture, and its ability to soak up bright dyes.

The spider makes silk for many uses, and gives it a different flavour in each case. The web is spun from dragline silk; other silks are used to make supporting fibres, threads that attach the web to the branch or the rafter, strands to bind prey, strands to swaddle the developing larvae, and so forth. All of these silks are composed of protein

15. Spider silk is one of the strongest known fibrous materials.

chains in which the amino acids glycine, alanine, and serine dominate. But the precise blend of components is tailored to the silk's use.

The wondrous properties of silk are a consequence of the way its protein chains are organized. Whereas in collagen and keratin the chains are coiled into helices, in silk the basic organized structural element is not a coil but a sheet. Neighbouring chains sit side by side in aligned ranks, and each chain is linked to those on either side via relatively weak hydrogen bonds (see page 41), which zip the chains together into beta sheets (Fig. 16).

The orderly, relatively rigid sheets can stack on top of one another to create tiny, three-dimensional protein crystallites. In silk fibres these crystallites are microscopic, extending perhaps only twenty-millionths of a millimetre in any direction. Beyond the crystallite regions, the protein chains continue into less orderly regions, where they are tangled together. So silk is really a kind of composite of tiny crystals dispersed in a more flexible protein matrix – a little like

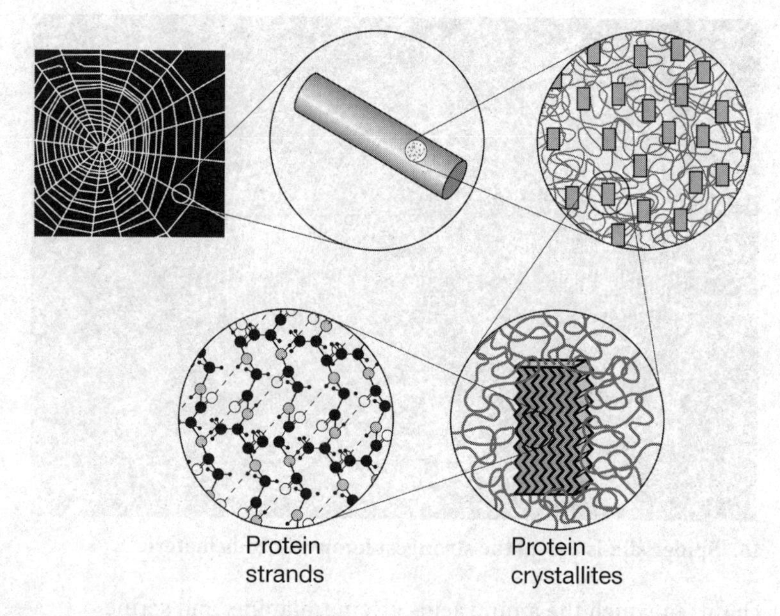

16. There are several levels to the hierarchical structure of silk. At the molecular scale it is organized into orderly (crystal-like) beta sheets of parallel strands. The dashed lines denote hydrogen bonds.

bone, except that the crystals are now not minerals but proteins themselves.

The crystalline regions of the silk protein generally have a regularly repeating sequence of amino acids. In the cocoon fibres of the silkworm *Bombyx mori*, for instance, the sequence glycine–alanine–glycine–alanine–glycine–serine repeats along the chains. In the disorderly regions, the sequence is irregular.

Silk fibres are insoluble in water – they would be of little use if they dissolved in the morning dew. Yet the spider spins the threads from a solution of protein molecules in water, thereby achieving the seemingly miraculous feat of turning a soluble molecule into an insoluble one. This change in solubility is a result of a change in the way that the chains are organized. The spider manufactures silk protein in its silk gland, at which point it is still soluble. This solution passes from the gland towards an outlet in the body called the spinneret. As it makes this journey, the silk solution loses water and becomes more concentrated, and the chains begin to get zipped together by hydrogen bonds. By the time the silk leaves the spinneret, most of the water has been squeezed out and the chains form beta sheets. Once the molecules are closely packed together in these crystalline regions, it is hard for water molecules to penetrate between them, and so the silk fibres are essentially solid and insoluble.

Molecular scientists cannot yet design artificial polymers with this sort of hierarchical structure, since it is very hard to control the way that molecules pack together over several different size scales at once. The molecular structure of synthetic polymers – which atoms they contain, and in what order and what spatial arrangement – can now be specified quite accurately; but it is another matter to 'program' these molecules to gather into particular kinds of cluster, or to cross-link them at definite locations.

Nevertheless, polymer chemists now know many tricks for engineering particular properties into their materials – making them hard or soft, say, or capable of forming strong fibres. In 1839 Charles Goodyear discovered how to cross-link the polymer molecules of the gum extracted from the tropical brazilwood tree *Hevea brasiliensis* by heating it with sulphur. This so-called vulcanization process converts the soft gum into the stretchy form we call rubber.

Rubber gum is made up mostly of a hydrocarbon polymer called polyisoprene, the chains of which are composed only of interlinked carbon atoms with some hydrogen atoms attached. Many of the plastics now manufactured on a large scale, such as polyethylene, polypropylene, and polystyrene, are also hydrocarbon polymers, made from the products of oil refining. But the rubber of truck tyres is natural, since it remains too hard to make it synthetically.

Until the mid-twentieth century the linking together of small hydrocarbon molecules to form synthetic polymer chains was a haphazard process, producing many different types of chain: some short, some long, some branched, some straight. Because the structure of the chains determines how they pack together, and because this packing determines the properties of the bulk material, this lack of control meant that polymer chemists could do little to fine-tune these properties. Whereas delicate structural control allows nature to make several apparently different fabrics from the same raw material, chemists just got the same kind of plastic every time. In the 1950s, special catalysts were developed that provided greater control over the molecular structure of the chains, and therefore over the way that they pack. This meant, for example, that polyethylene could be prepared in a new, hard, high-density form with a wider range of uses than the older, softer, low-density form – for example, as stiff drums and bottles for packaging, or as pipes and sheets in engineering and construction.

These two forms differ in the degree of orderliness with which the chains are packed together; that is to say, their degree of crystallinity. As silk demonstrates, greater crystallinity results in a tougher, stiffer material, as well as making it denser and less soluble. These are desirable attributes for strong fibres. The high strength arises because molecular chains that are packed in a closer, more orderly manner cohere more avidly.

So one of the principal challenges in making strong polymer fibres is to enhance the alignment of the molecules. Just about any polymer made of straight-chain (that is, non-branched) molecules has the potential to align its chains. But in silk this is helped by their tendency to 'zip' together to form crystalline beta sheets. In other words, the chains themselves have a kind of 'alignment instruction' programmed in. Can we make artificial polymers like this?

Indeed we can. The silk protein chains line up because of the forces of attraction between the units on different chains, which allow the chains to lock together like zips. Similar interactions exist between the chains of a class of synthetic polymers known as aramids, from which the renowned Kevlar fibres produced by DuPont are made.

Kevlar is one of the best candidates so far for tethering a Space Platform. It has a tensile strength greater than that of steel, but is much lighter. The fibres are used as the strengthening cord in rubber tyres, as a fabric for bulletproof clothing, as a reinforcing material in hard composites used for aerospace engineering, and even as woven cables for anchoring oil-drilling rigs.

But gram for gram, silk is stronger still. The high degree of chain alignment in a silk strand does not simply come from the fact that the molecules tend to zip together in solution. As the protein solution flows down a passageway from the spider's silk gland towards the spinneret, the polymer chains line up with the direction of liquid flow, in the same way that our hairs can be pulled into alignment by the wind. The same process operates also during extrusion of the thread from the spinneret. It is called shear induced alignment (since the flowing liquid experiences a so-called shear force).

Thus, even though it is possible to make silk-like proteins artificially, making silk threads from them is another matter entirely. If a fibre is mechanically extruded from a solution of (natural or artificial) silk protein just like pulling a thread from tacky glue, the fibre is still not as strong as real silk thread. Some researchers believe that only by building a tiny machine that mimics the spider's spinneret will we be able to make artificial silk that compares favourably with the natural material.

Shear-induced alignment is used in an industrial process to make extremely strong fibres from polyethylene. The fibres are drawn from a gel-like substance in a complex procedure that results in a very high degree of alignment of the polymer chains. These fibres, sometimes called 'rocket wire', are even stronger than Kevlar, and, unlike many organic materials, they are chemically very stable. This makes them well suited for use as long-term surgical sutures.

Material genes

One way of making artificial silk polymer is to import silk-making genes into bacteria using biotechnological techniques. Just like any other protein, silk is genetically encoded in DNA: the sequence of its amino acids along the chains is determined by a corresponding sequence of nucleotide units in a gene in the spider's chromosomes. In other words, the spider possesses a molecular blueprint for making the polymer. (But notice how, because of the complexities of the spinning process, this blueprint alone is not enough to make good silk threads!)

The ability of living organisms to define the molecular composition of a polymer with complete accuracy is an enviable one. Modern synthetic techniques allow chemists a great deal of control over the composition of a chain – for example, they can make hybrid polymers by grafting side chains of one chemical type onto a main chain of another, or by alternating between blocks of one type of unit and blocks of another along a single chain. But making a polymer perhaps thousands of units long in which many different units recur in a sequence of arbitrary complexity – and yet with every molecular chain in the material being identical – is something far beyond our current synthetic capabilities. In a linguistic analogy, our state-of-the-art polymers read something like this: aaaaaaabbbbbbbaaaaaaabbbbbbbaaa. . . Nature's polymers, meanwhile, are more like this entire sentence, and pregnant with meaning.

Biotechnology allows the genes from one organism to be snipped out of the respective stretch of DNA and pasted into the DNA of another organism. The recipient then (all being well) treats the new gene as if it were its own, and uses its transcription and translation machinery to make the corresponding protein. One of the potentially valuable and, I think, less controversial prospects of biotechnology is to transfer human genes into bacteria, which can then be bred in fermentation vats to make proteins needed for medicine. But some polymer scientists have realized that this is also a way to make protein-based materials.

As well as transferring real silk genes into bacteria,* one can 'write' artificial genes and express them this way. Structural proteins typically have repetitive amino-acid sequences, since the fibres they form are generally quite uniform along their length. It is relatively easy to make synthetic genes with repetitive sequences, since one need only make the individual repeat units and then join several of them together. (There are ways of ensuring that the resulting synthetic gene has only a specified number of copies of the repeat unit.) Using this idea, some researchers have begun to make genetically engineered protein-based materials in which the chains are programmed to adopt particular structures. 'Designed' silk-like materials have been made in this way, as well as protein-related synthetic materials similar to collagen and elastin, an elastic protein in skin. Hybrid materials can be envisaged that marry a natural protein (such as an enzyme) to a synthetic protein-like chain designed to behave as a material – for example, to pack into insoluble beta-sheet crystallites that will form a water-resistant surface coating. A possible attraction of protein-like materials for medical applications is that they would be biocompatible and biodegradable.

The cell's skeleton

Cells are laced with a network of fibres more rigid than the rope-like structural proteins collagen and keratin. These are called microtubules, and they provide the rod-like tracks on which molecular engines ferry packages around the cell. Microtubules also

* There is also ongoing work to obtain silk protein from goat's milk, by transferring silk genes into goats.

provide a scaffolding on which a cell alters its shape – for example, allowing an amoeba to extend a pseudopod. The hairlike cilia appendages that protrude from cells in our respiratory tract to push dirt-filtering mucus around, and the whiplike flagella that bacteria use to propel themselves with screwlike motions through a fluid – these too are microtubules.

As the name suggests, they are hollow, tubular structures. They are made up of a protein called tubulin, which is not fibrous but compact and globular. Tubulin consists of two nearly identical molecules, which pair up in a kind of dumbbell. The dumbbells stack together like bricks in a cylindrical chimney (Fig. 17).

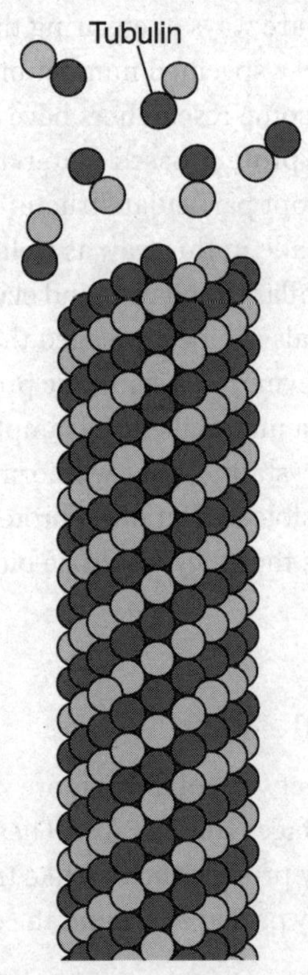

17. Tubular protein filaments called microtubules comprise the scaffolding of cells.

Tubulin units can attach themselves to and detach from the ends of microtubules, by which means the fibres are lengthened or shortened. An amoeba can retract its 'leg' simply by disassembling the microtubules that pushed it out. This ease of assembly enables microtubules to play a central role in cell division. Once the dividing cell has made copies of its chromosomes, it creates two clusters of radiating microtubules called asters. When the microtubules emerging from one aster meet those from the other, they merge at their ends, forming a bundle of bridging fibres between the focal points of the two asters, called the mitotic spindle (cell division is called mitosis). The chromosomes are attached to this spindle and are pulled apart so that half of each is dragged to either pole (Fig. 18). In this way the mitotic spindle provides a structure on which the duplicated genetic material can be separated out into two complete sets. The cell is then stretched and split into two halves on the framework of microtubules, each half containing a full complement of chromosomes.

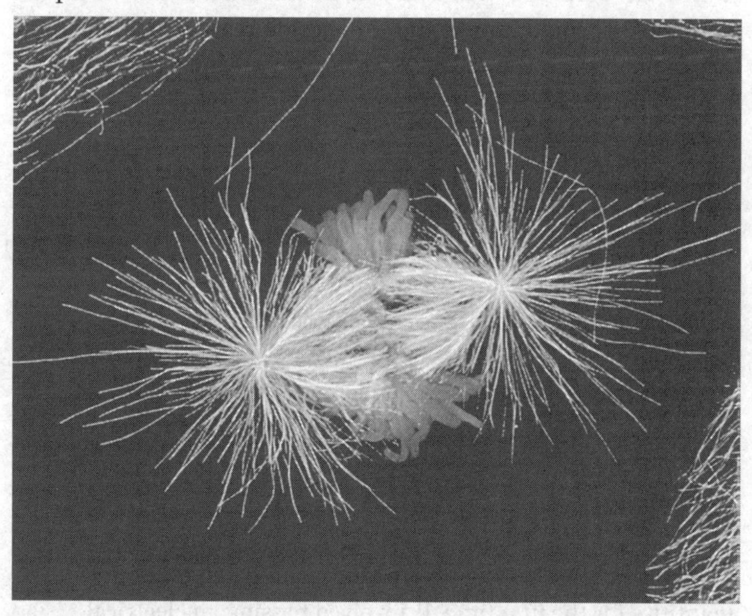

18. The mitotic spindle is composed of microtubules. It constitutes the framework on which chromosomes (visible here in the centre of the bundle) are arranged and sorted during cell division.

The anti-cancer drug taxol (see page 29) works by disrupting the assembly of the mitotic spindle and so arresting the proliferation of cancer cells. It prevents the disassembly of microtubules, which is essential as they search blindly for one another from the two asters. Unfortunately, taxol has this same effect on healthy cells; but, since cancer cells divide more rapidly, they are the hardest hit.

Carbon pipes

Until the 1990s, little serious thought had gone into the idea of making strong synthetic fibres from tubular molecules. It is easy enough to join small molecules together in chains, but tubes look much more difficult to orchestrate.

But in 1991 a Japanese microscopist named Sumio Iijima, working at the NEC Corporation in Tsukuba, discovered a kind of tubular molecule that could furnish the strongest fibres known. Called carbon nanotubes, these structures literally assemble themselves from a vapour of atomized carbon.

Iijima was investigating a technique used to make carbon molecules called fullerenes, in which dozens of carbon atoms are joined together in hollow cages. The first fullerene, a sixty-atom cage called buckminsterfullerene, was discovered in 1985. Its curious name derives from that of the US architect Richard Buckminster Fuller, who pioneered the 'geodesic dome' made from hexagonal and pentagonal facets. Buckminsterfullerene, or C_{60}, has this same structure at the molecular scale – hexagonal and pentagonal rings of six and five carbon atoms respectively, linked into a spherical cage (Fig. 19).

In 1990 a method was discovered for making fullerenes in large quantities for the first time. It involved passing an electrical discharge between two rods of graphite (which is pure carbon). The energy of the spark vaporized some of the graphite, and the carbon

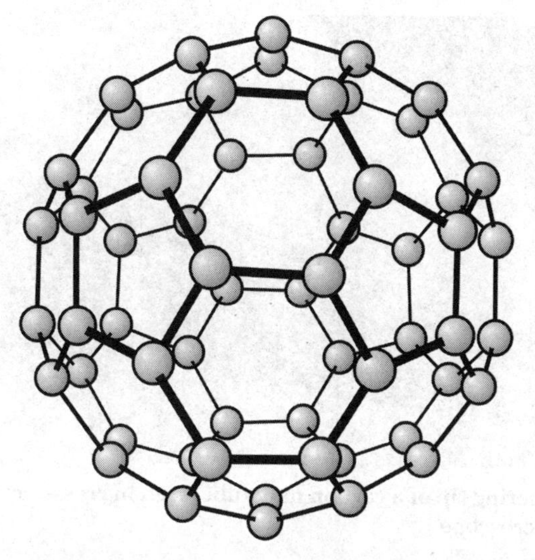

19. The C_{60} molecule is approximately spherical, and is composed of hexagonal and pentagonal rings of carbon atoms.

atoms joined together, as the vapour cooled, to form C_{60} and other carbon cages. But when Iijima used slightly different conditions in his fullerene generator, he found something new when he examined the sooty debris under the electron microscope.

The soot was full of needle-like objects, just a few nanometres in diameter. On closer inspection, Iijima saw that these were hollow cylinders of carbon, and that each one contained several cylinders nested inside one another like Russian dolls (Fig. 20). These objects became known as carbon nanotubes. No one had previously suspected, even after the discovery of fullerenes, that carbon atoms could arrange themselves spontaneously in this way.

The walls of nanotubes are made of pure carbon with the atoms arranged in sheets of hexagonal rings. This is the same structure as graphite, except that in nanotubes the sheets are curled up into cylinders. At the tube ends, the sheets are kinked into flat-faced end caps. We normally think of graphite as a weak material – after all, it is used in pencils because simply dragging it over paper rubs off

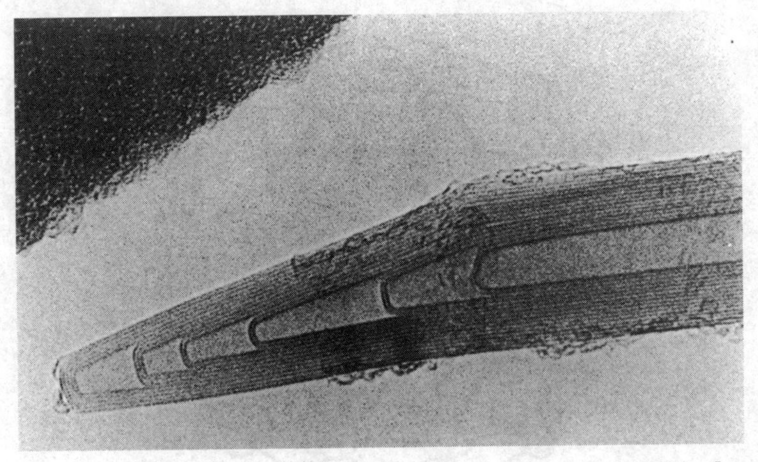

20. The tapering tip of a carbon nanotube, seen in cross-section in the electron microscope.

some of the black carbon. But this weakness is the result of the flat sheets of carbon being only loosely bound to one another, so that they can slide across each other. The individual sheets, in which the atoms are tightly bound, are predicted to be extremely strong – comparable, in fact, to diamond. When these graphite-like sheets are partially joined together by chemical bonds, as they are in conventional carbon fibres, the material gains a great deal in both strength and stiffness.

Carbon nanotubes are potentially the ultimate carbon fibres – the tubes are seamless sheets of graphite-like carbon. They are predicted to have a greater tensile strength than diamond; greater too than Kevlar, silk, or any other fibre natural or artificial you care to mention. Realizing this, Richard Smalley of Rice University in Houston, one of the discoverers of fullerenes, proposed that, if the Space Elevator were to be built, carbon nanotubes would hold it to the Earth.

But there is a snag. So far, carbon nanotubes cannot be grown to a length of more than a fraction of a millimetre. That is not a very long tether. It also makes measurements of the true strength difficult to conduct, although some experiments at the microscopic scale have already tentatively confirmed that these tubes are indeed very strong and stiff. (As strong and stiff as diamond? That is still uncertain.)

To make useful cables from carbon nanotubes, we will need to develop a way of controlling their growth so that the tubes can be extended indefinitely. In the synthesis of normal polymers, there is a technique called 'living polymerization' that allows the chain growth to be temporarily arrested and restarted at will, so that more and more units can be attached to ever-lengthening chains. If someone were to develop a similar process for carbon nanotubes, it would be revolutionary. But so far there is only a hazy understanding of how the tubes grow, making it hard to see how this could be controlled.* For the present, the Space Elevator will have to wait.

* A tentative first step has been reported by French researchers Philippe Poulin and colleagues at the University of Bordeaux I in Pessac. The team made nanotube fibres and ribbons by injecting nanotubes, suspended in water by the action of soaplike molecules, into a viscous liquid polymer. The nanotubes line up as they are injected from a capillary, in much the same way as silk proteins become aligned as they leave the spider's spinneret. The aligned tubes stick together in fibres that can be dried and handled. But these strands are not made from *continuous* nanotubes, and so they are not yet as strong as conventional carbon fibres, let alone diamond.

Chapter 4
The burning issue: molecules and energy

Imagine if the motor car was like the human body, so that its top speed was sustainable only for short sprints. No more driving down the motorway at (let's be honest) 80 miles an hour – that could only be managed for a half-mile trip to the shops. The further you had to go, the slower the speed, so as not to wear the poor vehicle out. (Perhaps you, like me, have driven cars that do seem to behave in this way?)

Superficially, cars and humans are more similar than one might think, and not for the reasons Flann O'Brien would suppose.* Both obtain energy by burning energy-rich fuel in oxygen; and both release an exhaust of carbon dioxide. Yet cars do not get tired when moving at close to top speed for long periods of time. Provided that you keep refilling the tank, they can keep it up almost indefinitely, whereas no sprinter could sustain their speed for an entire marathon even if they were continually chewing on glucose tablets. Why not?

The question leads us into the molecular mechanisms that power the body, which are known as metabolic processes. In many ways, it is metabolism and not replication that provides the best working definition of life. Evolutionary biologists would say

* If you don't know what I mean, read *The Third Policeman* and find out!

that we exist in order to reproduce – but we are not, even the most amorous of us, trying to reproduce all the time. Yet, if we stop metabolizing, even for a minute or two, we are done for.

Feel your hand. It is warm. (If it does not feel that way, try your armpit, or your tongue.) We are typically warmer than our surroundings. Whether waking or asleep, our bodies stay close to a healthy temperature of 37 °C. There is only one way of doing this: our cells are constantly pumping out heat, a by-product of metabolism. Heat is not really the point here – it is simply unavoidable, because all conversion of energy from one form to another squanders some of it this way. Our metabolic processes are primarily about making molecules. Cells cannot survive without constantly reinventing themselves: making new amino acids for proteins, new lipids for membranes, new nucleic acids so that they can divide. The wheels of the cell can never stop turning while we still live – and turning wheels consume energy.

So the community of the cell is rather like the stereotypical vision of a society during the Industrial Revolution: a culture of manufacture, in which a large part of the workforce is dedicated to generating energy. There are hundreds or even thousands of power plants in every cell of our liver, where energy production is paramount. Like the dark satanic mills of William Blake's *Jerusalem*, they produce waste as well as useful things; but cells have, on the whole, more efficient means of ensuring that they do not foul their own beds.

Since energy production is essential for life, a perusal of the machinations of metabolism can be a worrisome affair. We see what a fragile thing life is – for, if only this or that process were to be disrupted, the whole system would grind to a halt, just as our social structure hangs on the premiss of an uninterrupted electricity and gas supply (not to mention clean air and sufficient water). It is a testament to evolution's inventiveness that it has shaped a machine (if you will forgive an Enlightenment metaphor) so robust that it can run for eight decades or so, God willing, before its parts begin irreparably to fail.

Into the fire

But let us begin with some simplicity. Or at least, with the appearance of the same. In the 1850s, the British scientist Michael Faraday delivered at the Royal Institution in London a series of lectures on the 'Chemical History of a Candle'. He wished to show that in the incandescent glow of the candle's flame one could read the whole of chemical science as it was then understood – and more than just chemistry. 'There is no better, there is no more open door', he said, 'by which you can enter into the study of natural philosophy'.

To the trained eye, any natural phenomenon will serve as Blake's grain of sand, a window to the infinite universe. The candle will do, especially in nineteenth-century London, when showcase lectures did not have to be visually stunning. I do not really know the details of how a combustion engine works, but I know that it is not so very different from how a candle burns. It uses oxidation, the combination of some flammable fuel with oxygen in the air to produce heat and, in this case, light.

Even combustion of paraffin wax is still not understood in all its intricate details; certainly, there are many important aspects to the problem that Faraday did not know. Yet the essence of combustion is the essence of all chemical processes that generate energy. First of all, it is a downhill process.

That is the crucial point about chemical change, and about all other processes of change in the universe – they have a downhill and an uphill direction, and naturally enough they proceed in the downhill direction. What determines the topography? In the end, it is that now semi-mystical thing called entropy. Here is the Second Law of Thermodynamics, against which no one may appeal: in all processes of change, the total entropy of the universe increases.*

Popular culture equates entropy with disorder. It is not a bad shorthand: a system's entropy is a measure of how many ways its atoms can be reconfigured without any noticeable difference. If someone were to come into my office and reshuffle the chaos of paper into different heaps, the chances are I would not notice. But if they were to rearrange all the papers in the carefully ordered filing system that I can only dream about, I would soon detect the change. Crudely speaking, ordered systems have fewer indistinguishable configurations – a lower entropy – than disordered ones.

The Second Law is, then, really an expression of the following: it is more likely that a system will progress from a more-ordered to a less-ordered state, simply because there are more of the latter than the former. When one is dealing with systems that contain countless trillions of molecules, this probabilistic statement becomes a near certainty. The Second Law is a law only because its violation is overwhelmingly improbable.

But of course things *do* sometimes get more ordered. Water vapour condenses into symmetrical, six-pointed snowflakes. We can go out and arrange a pile of bricks into a house, and not be arrested for transgression against cosmic law. This is true, but does not violate the principle that entropy must increase. For the Second Law applies to the universe as a whole. Order can be bought at the price of a more-than-compensating increase in chaos elsewhere. Usually this compensation is paid in heat. Yes, we can build a wall by hard manual labour – but by the time we have finished, we will have

* Strictly speaking, this applies to irreversible change. Changes that can be reversed take place with zero change in entropy. By reversible, I do not mean putting something back where you took it from – all the body heat, all the movements of air, all the frictional rubbing, mean that picking up an object is irreversible.

radiated enough body heat to increase the disorderly thermal motions of our surroundings sufficiently for the entropy balance sheet still to be in credit. Order is earned through disorder.

Living cells maintain their organization apparently in the face of entropy's exigencies. This led the physicist Erwin Schrödinger to suppose, in his book *What is Life?*, that the answer lay in the somewhat nebulous concept of 'negative entropy', which living organisms extract from their surroundings. This sounds suspiciously like a kind of vitalism dressed up as thermodynamics; and one still encounters today talk of enzymes as 'anti-entropy devices'. Yet there is nothing special or mysterious about the mechanics of life. Feel that hand (armpit, tongue) again. What are you doing, but pumping out entropy into your environment?

I said that chemical change has a downhill direction – but sometimes it appears to go uphill. Enzymes are especially good at herding molecules uphill, although this can happen in the non-living world too. But in such cases, the reversal of direction is driven by some even more energetic downhill process. You can pull an object up a slope if it is coupled via a pulley to some more massive object moving downhill. The lighter weight rises as the heavier falls. Says chemist Peter Atkins, 'Understanding . . . biochemistry is essentially seeking heavy weights behind subtly hiding screens.'

Energy production in biochemistry is largely about pulling up as many weights as possible before a single heavy weight reaches the end of its fall. For animals like us, this heavy weight is the energy stored in the molecules we eat.

Alternatively, we might draw an analogy with harnessing a river for hydroelectric power. The water is going to go on falling whatever we do – it is never going to rise spontaneously back up the mountainside. The aim, then, is to capture as much of the energy of the cascade as possible. This is one of the major differences between our body's powerhouses and the candle's flame. In the flame, combustion rages uncontrolled, and all we get are heat and light. In the body, combustion takes place in a tightly controlled, graded sequence of steps, and some chemical energy is drawn off and stored at each stage.

Burnt sugar

A power station burns coal, oil, or gas – but it is much more than a flaming grate writ large. Burning is just a means to an end. The heat is used to turn water into steam; the pressure of the steam drives turbines; the turbines spin and send wire coils whirling in the arms of great magnets, which induces an electrical current in the wire. Energy is passed on, from chemical to heat to mechanical to electrical. And every plant has a barrage of regulatory and safety mechanisms. There are manual checks on pressure gauges and on the structural integrity of moving parts. Automatic sensors make the measurements. Failsafe devices avert catastrophic failure.

Energy generation in the cell is every bit as complicated. A brief description cannot hope to do justice to the awesome beauty of the system. The cell seems to have thought of everything, and has protein devices for fine-tuning it all.

The main inputs are fuel and oxygen: food and air. Starve our cells of oxygen and their flame goes out. We might delicately char the fuel before we swallow it, since these subtle beginnings of combustion create compounds that delight our palate. But everything from a chocolate bar to a spinach leaf to a pig's trotter must then be broken down to a more homogeneous fuel. This happens during digestion, when enzymes in the stomach and intestines break apart our culinary creations into their raw molecular ingredients.

There is a variety of energy-rich components in food, which can essentially be classified as either carbohydrates, fats, or proteins. Carbohydrates are polymers formed from molecules of the sugar glucose joined into long chains. Fats (also called lipids) may be broken down into so-called fatty acids and monoglyceride molecules such as glycerol during the digestive process. They yield twice as much energy as the same mass of carbohydrates, and the heart obtains 65 per cent of its energy from them.

Glucose, then, is one of the principal 'heavy weights' of metabolism. Its descent by enzyme-assisted combustion drives the formation of energy-rich molecules called adenosine triphosphate (ATP), which are used to power other processes in the cell. ATP is a biochemical power pack. A great many enzymatic reactions require ATP to drive them uphill.* ATP is the key to the maintenance of cellular integrity and organization, and so the cell puts a great deal of effort into making as much of it as possible from each molecule of glucose that it burns. About 40 per cent of the energy released by the combustion of food is conserved in ATP molecules.

ATP is rich in energy because it is like a coiled spring. It contains three phosphate groups, linked like so many train carriages. Each of these phosphate groups has a negative charge; this means that they repel one another. But because they are joined by chemical bonds, they cannot escape one another's baleful influence. Straining to get away, the phosphates pull an energetically powerful punch.†

The links between phosphates can be snipped in a reaction that involves water, for which reason it is called hydrolysis ('splitting with water'). Each time a bond is hydrolysed, energy is released. Setting free the outermost phosphate converts ATP to adenosine diphosphate (ADP); cleave the second phosphate and it becomes adenosine monophosphate (AMP). Both severances release comparable amounts of energy.

* ATP is not the only power source for the cell's machinery, but it is the most common. Some enzymatic reactions use other, similar energy-rich molecules, particularly guanosine triphosphate (GTP).

† Biochemical purists will note that this is a simplified argument for why ATP hydrolysis releases energy.

Good digestion

The business of breaking down food begins as soon as it enters the mouth, for saliva contains digestive enzymes called amylases, which start chopping up carbohydrate polymers into glucose. This is why food starts to taste sweeter when it has been chewed. In vegetables the digestible carbohydrate is largely starch; the cellulose of the plant cell walls is also a glucose polymer, but is resistant to the attack of amylases.

In the stomach, the food receives more severe treatment. Here, hydrochloric acid makes the gastric juices about as corrosive as battery acid. The acid loosens up the molecular coils of proteins in the food, ready for destruction by a gastric enzyme called pepsin.

The stomach is really just a storage chamber for undigested food, however. The disassembly process begins in earnest in the small intestine, into which the stomach releases its contents. The intestinal juices are spiced with a cocktail of small, hardy enzymes designed for specialized demolition jobs. While amylases attack carbohydrates, other enzymes break up the food proteins, fats, and nucleic acids. The fragments are absorbed through the intestinal lining and pass into blood and lymph vessels, which distribute these nutrients all around the body. The lining is covered in tiny folds and finger-like protrusions called microvilli, which vastly increase its surface area to ensure that nutrients are absorbed efficiently. Unfolded, the human small intestine would cover a tennis court.

Digestive enzymes are produced in the pancreas, a gland with ducts that empty into the small intestine. But these potent molecules are built to break apart the very molecules from which our cells are made – so how do they not destroy our own tissues?

The enzymes are manufactured with a molecular safety catch attached, which renders them inactive. In this form they are called zymogens. Not until they reach the intestine or stomach is the safety catch (a loop of polypeptide chain) removed, generally by another purpose-built enzyme. The digestive tract of the gut is coated with a layer of mucus, which protects it from digestion enzymes. If the protective coating becomes too thin, the corrosive juices go to work on the exposed tissues, causing an ulcer.

The contents of the stomach are broken down over several hours following a meal. But the body needs fuelling even after digestion is completed. So it stocks up reserves that can be called upon later. About one-tenth of the mass of our liver and 1 per cent of our muscle consists of a substance called glycogen, a compact, highly branched glucose polymer. The liver and muscle cells make glycogen from some of the sugar they receive, and store it in the form of small granules about one- to four-thousandths of a millimetre across – the pantries of the cell.

If the amount of sugar in the bloodstream falls below a certain level, the body knows it is time to start feasting on its glycogen reserves. Low sugar levels trigger the formation of two hormones in the pancreas – glucagon and epinephrine – which tell cells to start breaking apart their glygocen into glucose.

Too much sugar in the blood triggers a different warning system, inducing the formation of a hormone called insulin in the pancreas. Insulin is the signal for cells to start converting glucose to glycogen – to store it rather than burn it. As an indicator of a fuel glut, insulin also signals for cells to focus on manufacturing – for example, making proteins and fats (another emergency fuel reserve) – rather than energy generation.

Insulin is a peptide (protein-like) hormone, and is genetically encoded. A relatively common defect in the genetic message leads to the production of a precursory molecule to insulin (called proinsulin) that a manufacturing enzyme cannot convert to insulin itself in the normal manner. This failure to make insulin is one of the main causes of diabetes, and means that people with the disease have to administer regular doses of the hormone to regulate their blood sugar levels.

Wheels within wheels

Burning sugar is a two-stage process, beginning with its transformation to a molecule called pyruvate in a process known as glycolysis ('sugar-splitting'). This involves a sequence of ten enzyme-catalysed steps. The first five of these split glucose in half in an uphill process, powered by the consumption of ATP molecules: two of them are 'decharged' to ADP for every glucose molecule split. But the conversion of the fragments to pyruvate is a downhill affair that permits ATP to be recouped from ADP. Four ATP molecules are made this way, so that there is an overall gain of two ATP molecules per glucose molecule consumed. Thus glycolysis charges the cell's batteries.

Pyruvate then normally enters the second stage of the combustion process: the citric acid cycle, which requires oxygen. But if oxygen is scarce – that is, under anaerobic conditions – a contingency plan is enacted whereby pyruvate is instead converted to the molecule lactate.* This happens, for instance, if we are exercising so vigorously that the rate of oxygen supply is unable to keep pace with the rate of glycolysis, which the body steps up to meet the high energy demand. As lactate accumulates in muscle tissues it produces the painful symptoms of a 'stitch'.

Anaerobic metabolism is a relatively inefficient way of harvesting the energy of glucose. So extreme exercise leads quite quickly to muscle fatigue: energy is consumed faster than it can be generated, regardless of how much glucose is available. It is this that limits the

* There is so much going on in the body that one can rarely afford a general statement, and this is no exception; for some tissues convert substantial amounts of glucose to lactate even under aerobic conditions.

sprinter. The long-distance runner, meanwhile, finds a sustainable pace for which the full process of aerobic (oxygen-driven) metabolism can take its course by way of the citric acid cycle.*

This process is conducted in mitochondria, sausage-shaped compartments distributed many hundred to each human cell (Fig. 21). The first thing a mitochondrion does is convert pyruvate enzymatically to a molecule called acetyl coenzyme A (CoA). The breakdown of fatty acids and glycerides from fats also eventually generates acetyl CoA.

The cycle is a sequence of eight enzyme-catalysed reactions that transform acetyl CoA first to citric acid and then to various other molecules, ending with one called oxaloacetate. This end is a new beginning, for oxaloacetate reacts with acetyl CoA to make citric acid. In some of the steps of the cycle, carbon dioxide is generated as

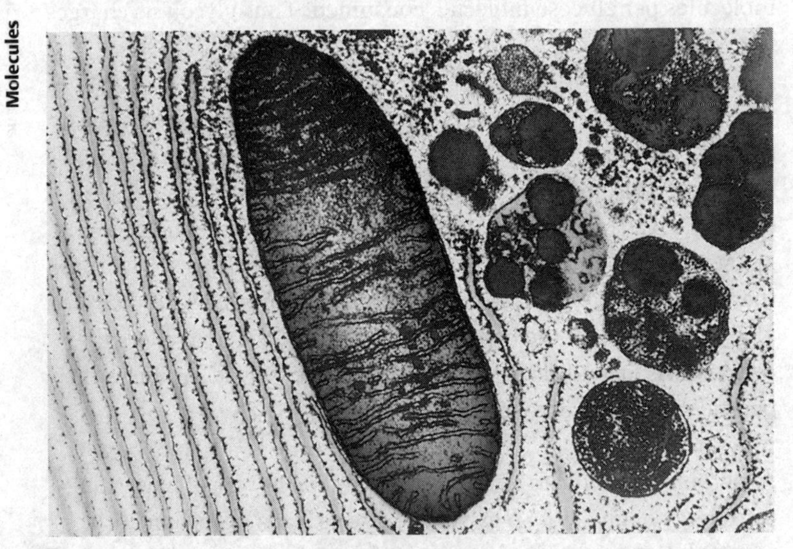

21. Mitochondria are self-contained compartments in cells. They are responsible for generating energy.

* Animals with higher metabolic rates can sustain exertion for longer periods, because they make ATP faster. The hummingbird can flap its wings furiously almost indefinitely, like an inexhaustible sprinter.

a by-product. It dissolves in the bloodstream and is carried off to the lungs to be exhaled. Thus in effect the carbon in the original glucose molecules is syphoned off into the end product carbon dioxide, completing the combustion process.

Also syphoned off from the cycle are electrons – crudely speaking, the citric acid cycle sends an electrical current to a different part of the mitochondrion. These electrons are used to convert oxygen molecules and positively charged hydrogen ions to water – an energy-releasing process. The energy is captured and used to make ATP in abundance.

The electrons do not flow as if down a metal wire, however; they are carried by a molecule called nicotinamide adenine dinucleotide (NAD). Two of the reactions in the citric acid cycle add an electron and a hydrogen atom to a positively charged ion of NAD, converting it to a molecule denoted NADH. This is the electron vehicle. Transfer of electrons from NADH molecules to an oxygen molecule initiates the formation of water from oxygen and hydrogen, while regenerating NAD. This is fed back into the citric acid cycle (Fig. 22).

There are further wheels within wheels. NADH does not give up its electrons directly to oxygen. Rather, this downhill process is conducted in several stages, and each one is tapped to charge up the mitochrondrion's batteries. The mitochrondrion is bounded by a relatively permeable outer membrane – through which come the raw ingredients that fuel the citric acid cycle – and an impermeable, highly convoluted inner membrane liberally studded with enzyme molecules. Within the inner membrane, NADH relinquishes its electrons to one of these membrane enzymes, whereupon the electrons are passed down a chain of other proteins. Eventually the electrons reach a membrane protein called cytochrome c oxidase, which has a site for binding oxygen molecules. It is this enzyme that finally transfers the electrons to oxygen.

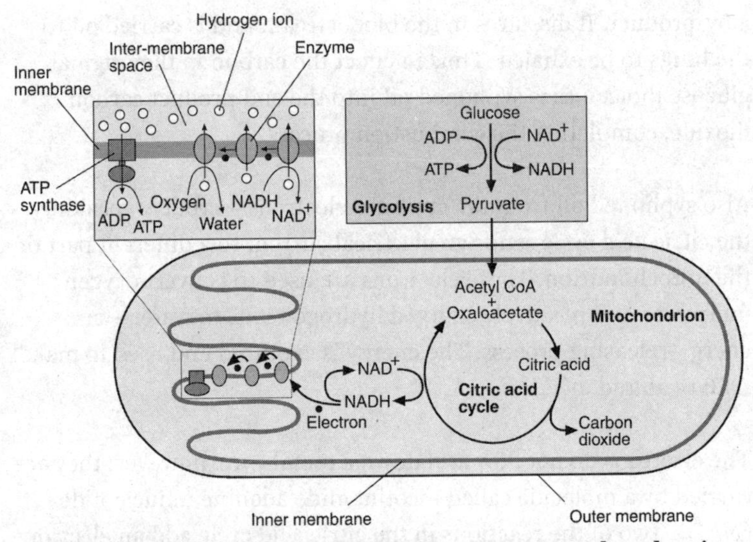

22. Energy production in the cell from burning sugars takes place in two steps: glycolysis and the citric acid cycle. The first is a linear sequence of reactions; the second, taking place inside the mitochondrion, can be regarded as a series of intermeshing cyclic processes.

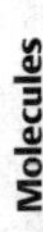

The oxygen-binding site of cytochrome c oxidase will bind certain other small molecules or ions even more strongly than oxygen – for example, cyanide and carbon monoxide. If this happens, the electrons can no longer reach the oxygen, and so this part of the machinery ceases to turn. Jamming of this cog causes the whole of the mitochondrion's mechanism to seize up – the citric acid cycle ceases – and the system is no longer able to produce ATP. The existing stocks are exhausted in minutes. So cyanide and carbon monoxide are potent poisons, which can cause rapid death by asphyxiation. The same applies for any substances that disrupt the proteins involved in the chain of electron transport from NADH to cytochrome c oxidase. Such substances include some of the most deadly poisons; this part of the cell's energy-generating machinery is particularly sensitive to attack.

The membrane proteins in the electron-transport chain pass on their bounty downhill to the next in line, and use some of the energy released to pump hydrogen ions from inside to outside the inner membrane. Thus, as the electron hops from one protein to the next, the concentration of hydrogen ions in the space between the inner and outer membrane increases.

This build-up is like the charging of a battery. The difference in hydrogen-ion concentration across the membrane, and the consequent difference in electrical charge of the two regions, is like the difference in electrical potential – the voltage drop – between a battery's terminals. Or you might think of the membrane proteins as pumps driving water back up a mountainside into a reservoir, from which energy can be extracted by letting the water run down again.

The hydrogen-ion reservoir drives ATP synthesis in the mitochondrion by powering a device very much like a miniature water wheel. The hydrogen ions flow back across the inner membrane through a membrane protein called ATP synthase, which harnesses the energy to make ATP from ADP (Fig. 22). ATP synthase has two main components. Its base, firmly lodged in the membrane, is a cylindrical channel through which hydrogen ions can flow. To the end of this channel, which opens onto the inside of the inner membrane, is attached a circular protein structure containing six globular subunits arranged in a ring. As the hydrogen ions pass, this circular head turns, and with each revolution it makes more ATP. ATP synthase is the nub of the cell's energy-generating machinery, and the eludication of its structure and much of its mechanism of action won Paul Boyer, John Walker, and Jens Skou the Nobel Prize for chemistry in 1999.

Glycolysis and the citric acid cycle are thus strikingly different kinds of metabolic process. One is anaerobic, the other aerobic. Glycolysis is potentially self-contained: one of its steps requires NAD, which is usually made in the citric acid cycle but which can be regenerated anaerobically instead by converting pyruvate to lactate. The two processes look very much like two independent kinds of metabolic pathway bolted together.

This is, in all probability, precisely what they are. The mitochondria are thought to have once been separate bacterial organisms – a notion supported by the fact that they possess their own DNA, distinct from the main genetic library in the cell's nucleus (see page 45). It is believed that, around two billion years ago, the mitochrondria entered a symbiotic (co-dependent) relationship with single-celled organisms that metabolized anaerobically using the glycolytic pathway.

Around this time, the spread of green algae (primitive plant life) led to a sharp increase in the amount of oxygen in the Earth's atmosphere; before then, there was rather little oxygen in the air. The mitochondria-like bacteria were able to 'breathe' oxygen using the citric acid cycle. Symbiosis between the anaerobic cells and the aerobes arose because the anaerobic cells made pyruvate, which the aerobes used for fuel, while the aerobes produced NAD to help drive the anaerobes' glycolytic process. Eventually, the anaerobes engulfed the aerobic organisms to become single, composite oxygen-breathing cells. The rest, as they say, is history.

Yeast never took this step: it is still a glycolytic anaerobe, living off sugar. But instead of turning pyruvate into lactate, yeast cells convert it to ethanol, to the delight of brewers for millennia. This is the process of fermentation, from which stemmed the whole of our understanding of enzyme action. It too generates carbon dioxide, the end product of combustion, which bubbles out of the brewing vats of the world.

Something in the air

Oxygen enters the body through the lungs. But it is not very soluble in blood, and so cannot be carried to the mitochondria simply by dissolving. (Carbon dioxide, in contrast, is soluble enough to make its own way from cells out to the lungs in the bloodstream.) Oxygen is carried by red blood cells packed with the protein molecule haemoglobin, which has a strong avidity for oxygen. Red blood cells could hardly be more single-mindedly designed for their job. Unlike other cells, they contain no DNA or RNA, no compartments and almost no other enzymes – they are basically sacks of haemoglobin. All the genetic and enzymatic machinery required for making this protein is ejected at the last moment as the cells mature.

The red colour of blood is akin to the red of rust, the hallmark of iron. At the heart of the haemoglobin molecule are iron atoms, caught in washer-shaped molecular traps called porphyrins. These iron-loaded porphyrins are known as haeme groups; they absorb blue-green light strongly and so look bright red. Each haemoglobin molecule contains four haeme groups, whose iron atoms provide attachment sites for oxygen molecules.

As an oxygen transporter, haemoglobin must be able to bind its cargo tightly, but also to let it go again. How can it do both? Haemoglobin employs a clever trick that is often used when a protein has to modify its behaviour once it has bound its target molecule. Put simply, one part of the protein can 'feel' what is going on in another part.

In the blood vessels of the lung, where oxygen is plentiful, the attachment of one oxygen molecule to a haeme group sends a shudder through the protein that encourages the other three haeme groups to take up oxygen too. Haemoglobin gives up its oxygen primarily in muscle tissue, where it transfers the molecule to myoglobin, another oxygen-binding protein, which has an even higher affinity for oxygen. Once one oxygen molecule has been taken from haemoglobin, a 'reverse shudder' weakens the grip of the other three haeme groups on their own cargo, and they release the oxygen more readily. These shudders – small but significant changes in the folded shape of the protein chain – are called allosteric motions.

All vertebrate animals and many invertebrates transport oxygen using haemoglobin, although the exact shape of the protein differs in different species. Arthropods and molluscs use a different oxygen-binding protein called haemerythrin, which also contains iron but not bound in a haeme group. This has a violet-pink colour when charged with oxygen, but is colourless when not. Some marine invertebrate creatures use a different metal for transporting oxygen: their oxygen-binding protein, called haemocyanin, is blue, betraying the presence of copper. They are the true blue-bloods of the ocean.

Leaf power

The oxygen-rich atmosphere of Earth is the work of plants. It is a by-product of photosynthesis, the biological use of the sun's energy to make molecules. Since plants and photosynthetic bacteria lie at the base of the food pyramid, all life on Earth is ultimately solar-powered. Without plants we are done for: we are given not our daily bread, and our cattle starve in barren pastures.

Photosynthesis is an old, old process. Algae were photosynthesizing at least three and a half billion years ago, when the continents were newly formed and ungreened. These organisms were the first *autotrophs*: 'self-feeders', making their own molecules from little more than light, water, and carbon in the air. About 60 billion tons of carbon are plucked every year from the carbon dioxide in the atmosphere and are turned into energy-rich biomass. Some of this we eat, burn, arrange into houses and tables, feed to livestock, pulp into paper, spin into cloth. Much of it falls, is broken down by micro-organisms, and is released back into the atmosphere as volatile carbon compounds. Over geological time, some will be buried, compressed into coal, or decomposed into oil or gas.

Photosynthesis depends on molecules that interact with light, absorbing some of its energy and channelling it into chemical processes. While mammalian cells have fuel-burning factories in

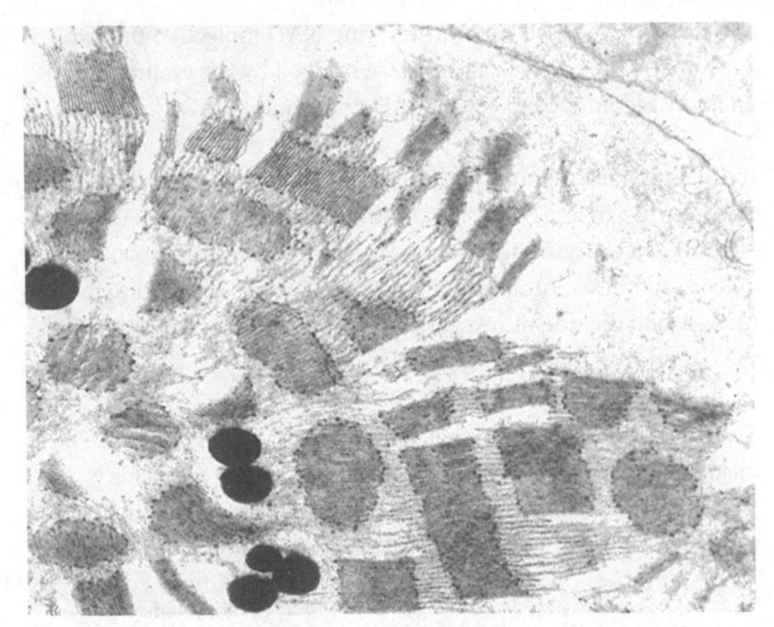

23. Plants capture sunlight and convert it to chemical energy in chloroplasts, filled with the stacked sheets of thylakoid membranes.

the form of mitochondria, the solar-power centres in the cells of plant leaves are compartments called chloroplasts (Fig. 23). The process conducted herein is, broadly speaking, the reverse of glucose metabolism. The chloroplast takes carbon dioxide and water, and from them constructs the sugar. 'Burning' glucose is an energetically downhill process, so it follows that the manufacture of glucose in photosynthesis is uphill. This is why the plant needs the energy of light rays to do it. Yet the plant uses this energy not just to create glucose for weaving into the cellulose walls of its cells, but also – and just as importantly – for making ATP molecules to drive the cells' chemistry.

There are several similarities between the processes of aerobic metabolism and photosynthesis. Both consist of two distinct sub-processes with separate evolutionary origins: a linear sequence of reactions coupled to a cyclic sequence that regenerates the molecules they both need. The bridge between glycolysis and the citric acid cycle is the electron-ferrying NAD molecule; the two sub-processes of photosynthesis are bridged by the cycling of an almost identical molecule, NAD phosphate (NADP).

In the first part of photosynthesis, light is used to convert NADP to an electron carrier (NADPH) and to transform ADP to ATP. This is effectively a charging-up process that primes the chloroplast for glucose synthesis. In the second part, ATP and NADPH are used to turn carbon dioxide into sugar, in a cyclic sequence of steps called the Calvin–Benson cycle (Fig. 24).

The first process takes place at the surface of a folded membrane inside the chloroplast, called the thylakoid membrane. This is studded with clusters of molecules called 'photosystems', in which light-absorbing molecules called photopigments initiate

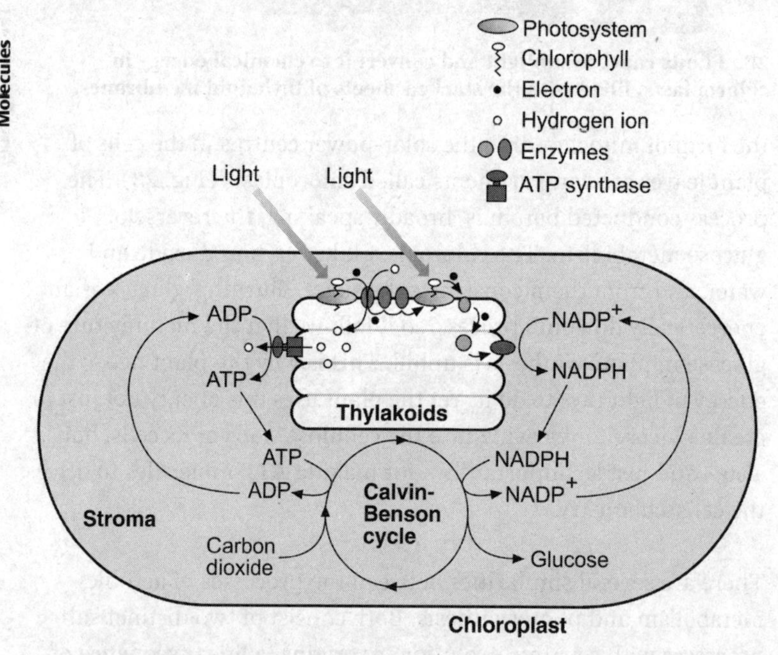

24. During photosynthesis, captured solar energy is used to split water and to make ATP, which then drives the conversion of carbon dioxide to glucose.

light-powered reactions. At the heart of the photosystems, called the photosynthetic reaction centre, is a molecule called chlorophyll *a*. This absorbs red and blue light strongly, and is thus responsible for the green colour of leaves.

When chlorophyll receives light energy, it becomes 'excited', like an apple tree that gets shaken. In its excited state, chlorophyll is less able to hold on to its outer electrons, and one of them comes free. The electron is passed on to an enzyme; once the enzyme has gained two electrons from 'shaken' chlorophylls, it can transform a positively charged ion of NADP to NADPH. The electron-deficient chlorophylls are then replenished with electrons plucked from water molecules in another light-powered reaction. The water is broken into hydrogen ions and oxygen atoms. The oxygen atoms combine into two-atom oxygen molecules, which the plant releases through openings in the leaf surface.

The electron taken from water is passed to chlorophyll along a chain of molecules embedded in the thylakoid membrane. Each transfer step is a downhill process that releases energy, some of which is tapped to pump hydrogen ions into the inner space of the thylakoid membrane. This imbalance is then harnessed as an energy source by ATP synthase molecules lodged in the membrane, which perform their windmilling conversion of ADP to ATP.

But the task of photosynthesis is not finished with water-splitting and the production of the energy source ATP and the electron source NADPH. These two ingredients are released into the fluid outside the thylakoid membrane, called the stroma, where they drive the reactions of the Calvin–Benson cycle, which turn carbon dioxide to sugar. These processes are called the 'dark reactions', because they do not require light directly. In 1961 US chemist Melvin Calvin won the Nobel Prize for deducing most of this sequence.

Chemists are currently interested in designing artificial molecular systems that, like chloroplasts, harness sunlight to drive chemical synthesis. A team at Arizona State University has mimicked the chloroplast in synthetic cell-like (and cell-sized) structures called liposomes: hollow, spherical membranes made from lipid molecules. The researchers peppered liposome membranes with designed molecular assemblies that perform the same task as photosynthetic reaction centres – using light energy to pump hydrogen ions into the liposome's hollow interior.

The researchers inserted molecules of ATP synthase into their liposomes, which release the hydrogen ions and make ATP into the bargain. The hope is that the energy stored in ATP can be tapped for chemical synthesis – for example, to conduct some industrially useful biochemical reaction.

Ending with a bang

Glucose and candle wax (a hydrocarbon) are both 'embodied energy': breaking their bonds with oxygen releases energy as heat (unless channelled into other forms). But some chemists seek molecules that pack a bigger punch. How much energy can be crammed into a molecule?

Alfred Nobel occupied himself with that question in the nineteenth century, and the resulting irony – the fortune amassed from the invention of dynamite, which now funds an annual peace prize – is well known. Nobel's innovation was the invention not of an energy-rich molecule but of a means for packaging an existing explosive into a form that was less likely to blow up in one's face.

The oldest explosive is gunpowder, a mixture of sulphur, nitre (potassium nitrate), and charcoal devised in China around the eleventh century AD. This was deployed in the West with little change (and terrible effect) until the nineteenth century, when scientists began to uncover ways of making a bigger bang. In 1845 the Swiss chemist Christian Schönbein discovered nitrocellulose, a compound made by treating cotton fibres (cellulose) with nitric and sulphuric acid. This was the first 'semi-synthetic' polymer, a product both of nature and of the chemist's art. It was later developed into celluloid, a hard plastic, and rayon, the first 'artificial silk'. But nitrocellulose was also explosive, and became known as gun cotton.

Gun cotton's violent nature was hard to control; early attempts to manufacture it led to several deaths. In 1847 the Italian chemist Ascanio Sobrero synthesized a similarly hazardous substance called nitroglycerine. Alfred Nobel began to study this compound in 1859, attempting to find a way of rendering it stable until deliberately detonated. Persisting despite an explosion in 1864 that killed his younger brother, Nobel found that mixing nitroglycerine with a clay called kieselguhr produced a putty-like explosive that could be handled safely. He called it dynamite. In 1875 Nobel introduced gelignite or 'blasting gelatine', a jelly-like mixture of nitroglycerine and gun cotton, which was more powerful than either substance on its own.

These explosives were used largely for mining and construction blasting, and they made Nobel's fortune. But inevitably they were adopted for military use too. The British and French armies of the late nineteenth century used cordite, an explosive similar to blasting gelatine; the German troops took to trinitrotoluene (TNT), which explodes only if detonated with a secondary explosive. In *Brave New World* Aldous Huxley tells us of the rare chemistry of this lethal combination:

> $CH_3C_6H_2(NO_2)_3 + Hg(CNO)_2$ = well, what? An enormous hole in the ground, a pile of masonry, some bits of flesh and mucus, a foot, with the boot still on it, flying through the air and landing, flop, in the middle of the geraniums . . .

All of these compounds are organic substances containing the nitro group: a nitrogen atom with two oxygen atoms attached. The explosiveness of nitro compounds derives from the fact that nitrogen atoms recombine to form nitrogen molecules when the materials are ignited. These molecules have very stable bonds whose formation releases a lot of energy. The oxygen atoms, meanwhile, stimulate the combustion process, allowing it to happen very quickly. Developing more powerful explosives has been largely a matter of finding ways to pack more nitro groups into an organic compound. RDX or cyclonite, an explosive used in today's weaponry, improves on TNT in this respect. The most powerful explosive in current production is a nitrogen-rich compound called HMX, an abbreviation for 'high-melting explosive'.

In early 2000, a potentially still more energy-packed nitro compound was devised by chemists at the University of Chicago. Called octanitrocubane, it consists of a cube of eight carbon atoms, to each of which is attached a nitro group. Not only is this molecule extremely nitro-rich, but the cubic shape means that the bonds between carbon atoms are highly strained and easy to burst open. And the compact molecules should be able to stack together into a very densely packed crystalline form. The initial experiments have not yielded this high-density form, but calculations predict that, if it can be made, it will have an explosive energy content greater, gram for gram, than any known non-nuclear explosive.

Alfred Nobel's prize-giving bequest seems to have been a gesture provoked by the regret he felt for the lethal and destructive applications of his work. Clearly, not all chemists will feel this way about weapons-related research. My own view, for what it is worth, is that such work is arguably a form of scientific misconduct. But, more importantly, work on high explosives demonstrates that scientific research does not split cleanly into an amoral 'pure' science and a dirty and socially accountable 'applied' science or technology. Making octanitrocubane was a feat of technical brilliance; it was also funded by the US Defense Department.

Explosives reveal the Janus face of molecular science. They are part of its fun – for how many budding chemists have never sought out the classic exploding reactions one can conduct in a school lab? I know of at least one respected scientist who was expelled from his school for almost destroying it. (He now studies planet-sterilizing giant meteorite impacts.) Yet it is but a small step from these bangs and flashes to the destruction of Dresden and Hamburg. It is a perilous step.

Chapter 5
Good little movers: molecular motors

After-dinner speeches are not normally notable for launching revolutions. But Richard Feynman, who was engaged to address the West Coast section of the American Physical Society in 1959, was not a normal physicist. One of the most creative scientific minds of the post-war twentieth century, he is most vividly remembered by the world at large as a bongo player, a practical joker, a safe-cracker, the trickster figure of modern science.

Feynman's talk in 1959 was high-spirited but ultimately serious in its intent. He called it 'There's Plenty of Room at the Bottom', and it was about engineering on scales too tiny to see. 'What I want to talk about', he said, 'is the problem of manipulating and controlling things on a small scale'. By 'small', said Feynman, he did not mean 'electric motors that are the size of the nail on your small finger'. He meant small as in atoms.

'Imagine', he went on, 'that we could arrange atoms one by one, just as we want them'. This, he saw, is essentially what the chemist tries to do:

> The chemist does a mysterious thing when he wants to make a molecule. He sees that he has got that ring, so he mixes this and that, and he shakes it, and he fiddles around. And, at the end of a difficult process, he usually does succeed in synthesizing what he wants.

You can see that the physicist's view of what chemists do is hardly more sophisticated than a lay person's. But Feynman's description is really not so different from the one Primo Levi gives when explaining how chemists build molecules as engineers build bridges (see page 24). However, the chemist is traditionally accustomed to regarding his molecule as a substance, something to crystallize and put in a bottle. The physicist, on the other hand, sees it as a construct, like an engine component.

Feynman was essentially wondering whether physicists might figure out how to do what chemists do, but wearing an engineer's hat. Can we build a molecule by pushing atoms into place, one by one? In 1959 such a thing was unthinkable to anyone but a conjuror of the imagination like Feynman.

Yet he was not simply speculating idly. Even at that time, it was clear that technology was getting smaller. The invention of the transistor in the 1940s had shrunk the scale of electronics. Bulky boxes filled with vacuum tubes had been replaced by compact devices containing 'solid-state' circuits made from silicon transistors. The portable transistor radio was on every American beach. Engineers were becoming increasingly skilled at making tiny machine components – much more so, in fact, than Feynman realized. Hoping to provide some small impetus for driving miniaturization technology forward, he offered two prizes of a thousand dollars, to be funded by himself: one for making an electric motor measuring no more than 1/64th of an inch on any side, the other for writing the information from a page of a book in an area scaled down by a factor of 1/25,000. Feynman presumably anticipated that his money would be safe for some years to come – he did not imagine that someone (an engineer named William McLellan) would meet his first challenge within a few months.

Today we can go further. Tiny cogs and motors a tenth of a millimetre across have been carved out of silicon wafers using acid etching or electron beams (Fig. 25). But carving out parts from slabs of material is all very well until you reach a scale of around a tenth of a micrometre: current methods for making integrated circuits in silicon can just about make wires this thin. Beyond that these methods cannot go – it becomes like trying to split a human hair with a bread knife.

Researchers are starting to ask whether this 'top-down' approach still makes sense at such scales. Components this small are closer in size to molecules (medium-sized molecules are a hundred times smaller) than to silicon wafers you can hold and see in your hand. Should we then start making things from the bottom up – from single molecules?

Primo Levi confessed in *The Monkey's Wrench* that chemists fantasize about a tool kit for molecular-scale construction:

> we don't have those tweezers we often dream of at night, the way a thirsty man dreams of springs, that would allow us to pick up a segment, hold it firm and straight, and paste it in the right direction on the segment that has already been assembled. If we had those tweezers (and it's possible that, one day, we will), we would have managed to create some lovely things that so far only the Almighty has made, for example, to assemble – perhaps not a frog or a dragonfly – but at least a microbe or the spore of a mold.

Feynman too found inspiration in the molecular devices and artefacts of biology: 'in which chemical forces are used in a repetitious fashion to produce all kinds of weird effects (one of which is the author).' He realized that there are already molecular machines in biology. In 1959 this, had it reached the ears of biologists, would doubtless have been dismissed as the attempt of some foolish physicist to impose his own perspective on a field he clearly knew nothing about. But today biologists are quite happy to talk about proteins as molecular machines.

25. A micromotor carved from a silicon wafer.

This chapter looks at some of the most remarkable of these: protein molecules that create movement. They are molecular motors, often called motor proteins. The mini-motor that won Feynman's prize is a clumsy gargantuan in comparison, as a lumbering Diplodocus is to the nimble flea. The biological importance of motor proteins is immeasurable. Without them, we could not move a muscle; no birds would cross the sky, no fish ply the seas. Even bacteria would be immobile. But worse: cells could not divide, so there would be no reproduction. Without molecules to drive movement, there is no life.

Yet for the mechanic of the molecular world, motor proteins say something else. They show that molecular-scale engineering is possible: that we can scale down ideas familiar from the everyday world to the realm of molecules. Motor proteins are not unique in this respect, but they make the point with rare explicitness. I will describe how we might achieve similar goals from scratch, by making our own custom-built molecular motors. This leads us into the arena towards which Richard Feynman's talk was the first clear signpost: the science of *nanotechnology*, which is technology on the scale of nanometres – distances one can measure in molecules.

Front crawl

The shape of a molecule is never fixed: it is always vibrating and waggling its loose, floppy parts. Mechanical motion is ubiquitous in the molecular world.

Yet, in general, either molecular motions are random, like the meandering wriggle of a polymer chain floating in solution, or they average to zero, like the back-and-forth vibrations of a chemical bond. What we need from a genuine motor, in contrast, is motion with a directional bias – what one might call purposeful motion.

Any motor consumes fuel. You could regard this as the inevitable price of orderly motion, a toll imposed by the Second Law of Thermodynamics. Random molecular motion, on the other hand, can be had 'for free' – it is the incoherent molecular jiggle that is heat.

Our bodies conduct many types of directional transport. For example, the motion of cilia – hairlike appendages that line the air passages of our lungs and windpipe – moves a layer of mucus from the lung lining up to our throats, where it accumulates as phlegm. This mucus captures dirt, and so its export keeps the lungs clean. To move the mucus up the windpipe, the cilia cannot just thrash around but have to execute a coordinated sequence of movements, like the arms of a swimmer. They make a whiplike 'power stroke' followed by a slow, crawling 'recovery stroke'. Some single-celled organisms called protists do in fact use cilia on their cell surface to swim through water.

The molecular engine that drives these motions is a protein called dynein. Each cilium contains microtubules (see page 63) arranged around the circumference of a tube called the axoneme. The tubules are joined together in pairs called doublets, like the barrel of a twin-barrelled shotgun. There are nine paired microtubules per axoneme, and they are interconnected by dynein molecules, protruding at regular intervals like the legs of a millipede (Fig. 26).

To move the cilium, the microtubules walk over one another. Each dynein molecule has a 'leg' that bends by undergoing a reaction that consumes ATP. Dynein is basically an enzyme that breaks apart ATP and changes shape as a result. Calcium ions are also needed to

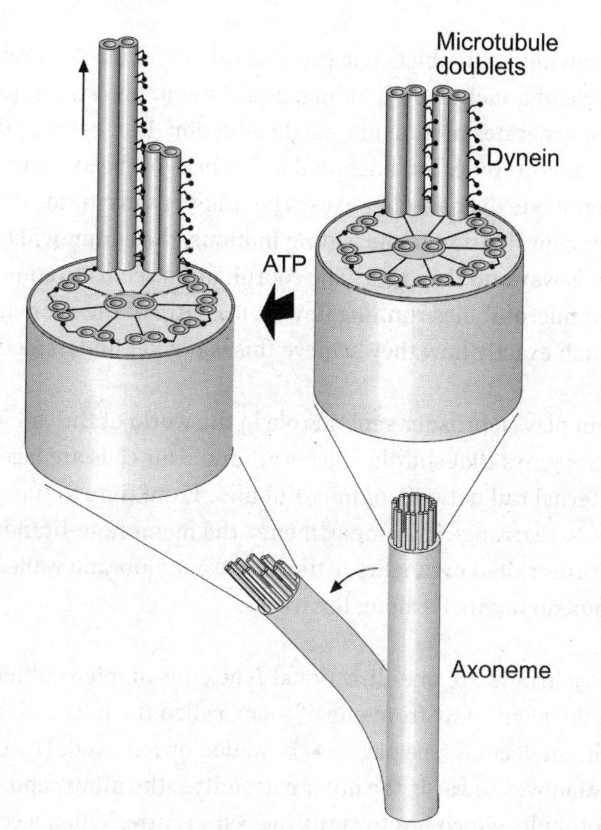

26. The bending of cilia is driven by dynein molecular motors.

trigger this reaction. The motion is controlled by nerve signals that trigger the injection of calcium into the cilium.

Since the dynein molecules all point in the same direction, they pull one microtubule doublet over another when they bend. If the molecules were then simply to straighten again, the microtubules would return to their original positions. In order to generate forward motion, each dynein molecule detaches itself from the second microtubule before straightening up, and then reattaches for the next 'power stroke'. Only when its 'foot' is attached to another microtubule can dynein break down ATP and switch to the bent state.

The crawling of doublets one over the other is thus like a ratchet: the cycle of attachment, bending, detachment, and straightening of dynein generates motion in a single direction. But, because the ends of the microtubules are anchored at the base of the axoneme, this sliding of one over another causes bending of the cilium. With proper coordination of the sliding motions, the cilium will bend first this way and then that. The coordination seems to come from a pair of microtubules running down the centre of the axoneme, although exactly how they achieve this is not yet understood.

Dynein plays a broader general role in the world of the cell: it is one of the engines that shuttle objects around. Our cells are laced with an internal rail network of microtubules. From time to time the cell needs to rearrange its compartments, the membrane-branded structures called organelles. Attached to a membrane wall, dynein can pull an organelle along the tracks.

These journeys are one-directional. The ends of microtubules are not equivalent: only from one of them, called the plus end, can tubulin molecules (see page 64) be added or removed. Dynein always moves towards the other extremity – the minus end – of a microtubule, which lies towards the cell's centre. When a cell divides in two, dynein pulls the duplicated sets of chromosomes along the microtubules of the mitotic spindle (see page 65), carrying them towards the centres of the respective nascent daughter cells.

For transport along microtubules in the other (plus) direction, a different motor protein called kinesin is used. Kinesin is perhaps the most anthropomorphic of molecules that induce motion, since it has two 'legs' and executes a waddling 'walk' in comparison with dynein's one-legged inchworm crawl. Kinesin too is powered by a reaction that consumes ATP and alters the protein's shape.

Kinesin is the cell's postman, delivering parcels from one organelle to another. For example, proteins must be sent from their point of manufacture (the endoplasmic reticulum) to the parts of the cell where they are needed. They are packaged inside little membrane spheres called transport vesicles, and kinesin carries a transport vesicle along the microtubule network to the right address.

Muscle power

Our own walking is enabled by muscle contractions and extensions. We are sprung and countersprung with so-called skeletal muscles, which, simply by shortening and relaxing, can control everything from a pianist's delicate finger movements to the pounding of an athlete's thighs.

Skeletal muscle is one of nature's hierarchical molecular materials (see page 53). It is a fibrous composite of bundles within bundles within bundles. Individual muscle cells are extremely elongated and enclose many-stranded cables woven from threads called myofibrils. Within the complex molecular substructure of these strands reside the secrets of muscle contraction.

Seen through the microscope, a myofibril is punctuated by light and dark bands of various widths. These give skeletal muscle a striped appearance at high magnification, which is why it is also known as striated muscle. The sequence of bands repeats periodically along the myofibril strand, and one repeat unit is called a sarcomere. The various bands within a sarcomere are given the kind of anodyne names – A band, H zone, and so forth – that are always telltale indications that no one had the faintest idea, when they were first observed, what they signified.

Noticing in the 1950s that these bands changed width when muscle contracted, Andrew Huxley, Hugh Huxley, and their co-workers proposed the so-called sliding filament theory of muscle action. Their idea was that the myofibril contains toothbrush-like structures facing one another and pushed together so that their bristles interpenetrate. Each sarcomere contains a double set of these pairs of brushes, placed back to back. The dark bands correspond to regions where the bristles interdigitate (creating a high density of molecules), whereas in light bands there is only one set of bristles (Fig. 27). The Huxleys suggested that the myofibril shortens – and so the muscle contracts – by deeper interpenetration of the sarcomere's bristles.

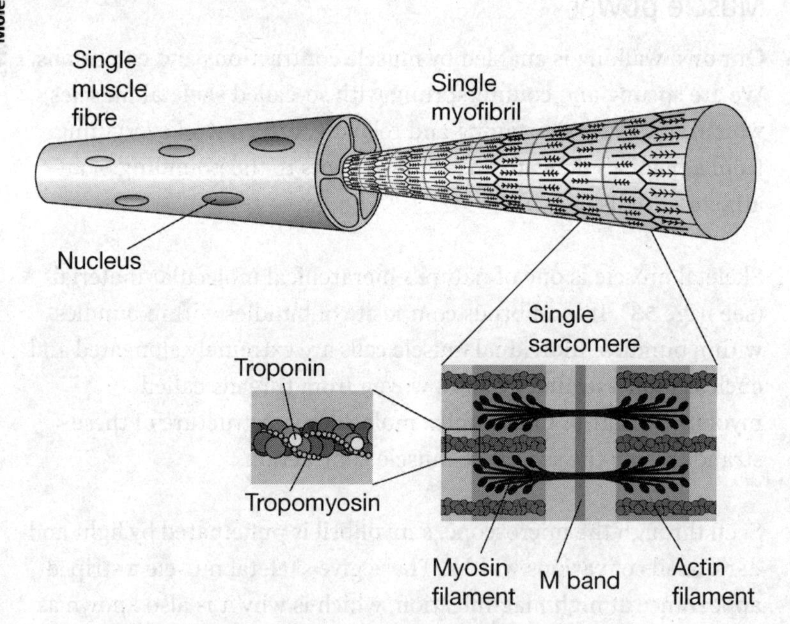

27. Interdigitating filaments in muscle allow it to contract.

This movement of filaments over one another is driven by the motor protein myosin, a long thin protein in which two helical chains twist around one another. At either end, the chains terminate in a pear-shaped head. Myosin molecules are gathered into bundles called myosin filaments. Each end of a filament bristles with myosin heads, like a head of corn budding from a sheaf.

Penetrating between the myosin bundles are filaments of a protein called actin. This protein is in fact globular in shape, but the globules link together to form a chain like beads on a necklace. In an actin filament, two chains of actin 'beads' wrap around each other in yet another double helix. The necklace is further adorned by strands of the protein tropomyosin, which wind their way along the actin filament. And at regular intervals sits a globular protein called troponin (see Fig. 27).

Muscle contracts when myosin heads attach themselves to the actin filaments and pull themselves along. The principle is the same as that by which dynein and kinesin move along microtubules: motion is generated by a change in shape of the attached motor protein, driven by the breaking-down of ATP to ADP. The myosin head swings on a hinge connecting it to the rest of the molecule. It kinks, detaches itself from actin, unkinks, and reattaches, and thereby ratchets along the actin filament in a series of power strokes (Fig. 28).

The whole process is under voluntary control, set in motion when a nerve impulse from the brain tells the muscle to tighten or relax. The tropomyosin and troponin proteins on the actin filament provide the switch. Muscles are wired to the brain by nerve cells called motor neurons, which are like wires biochemically 'soldered' to the outside of a muscle fibre. An electrical signal arriving at the end of the motor neuron triggers the release of calcium ions from a web of tubes – the sarcoplasmic reticulum – that extends through the spaces between myofibrils inside the muscle fibre. These calcium ions are captured by

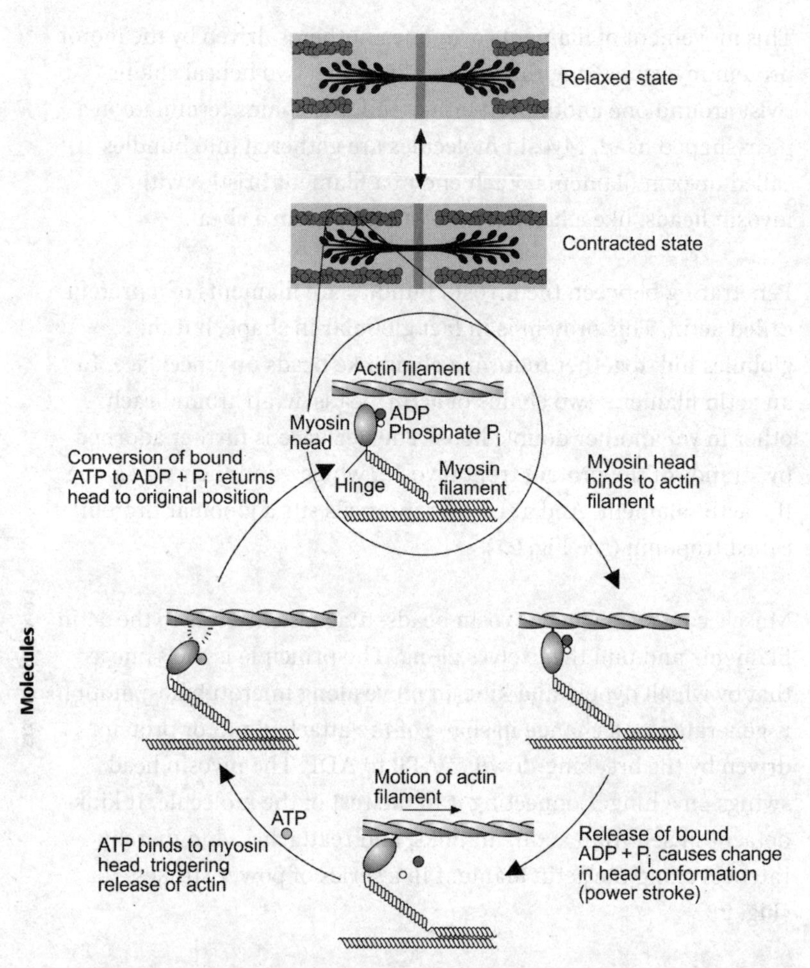

28. Muscle contraction is caused by the movement of myosin molecular motors along filaments of the protein actin.

troponin molecules on the actin filament, prompting the proteins to change shape. This in turn pulls on the tropomyosin strands, which twists the double-helical actin necklace and rotates the actin beads. It is this rotation that exposes the sites on actin to which myosin binds. The entire structure thus contains an elegant molecular transmission mechanism for switching contraction on and off.

Molecular tweezers

Once upon a time molecular scientists had to deduce all they knew about molecules from measurements made on many billions of them simultaneously. This can be a risky business, since we cannot always be sure how such measurements are related to the properties of individual molecules, just as the noise that emanates from a football stadium or theatre hall reveals nothing of the individual conversations people are having. But advances in experimental techniques that enable studies of single molecules – what they look like, how they interact, how they move – have over the past two decades opened up an entirely new realm of molecular studies. We are starting to get to know molecules in person.

One of the critical innovations is the invention of tweezers for molecular manipulation – the very tool that Primo Levi desired. The most remarkable thing about these tweezers is not that they are so fine but that they are literally intangible – they are made of light. They are called optical tweezers, and they trap objects in a very intense light beam. They allow researchers to address the kinds of questions one might ask about everyday mechanical motors: how efficient are they, how much load can they bear, how fast do they move?

Interaction between light and the electrons in molecules can create a force – a kind of 'light pressure' – on an object. If the object is small enough and the light intense enough, the object can be moved by this force. In optical tweezers, the intersection of two or more laser beams sets up a spot of extremely bright light. A small object within this bright pool experiences a light pressure from all sides that prevents it from moving in any direction. It is caught in an optical trap between the tweezers of the laser beams. If the beams are moved, the object is pulled along with them.

The force generated by a single motor protein can be measured by tethering either the motor or the object it moves along (an actin filament, say) to a microscopic plastic bead clamped between optical tweezers. Motion generated by the motor tugs the bead away from the centre of the trap by an amount proportional to the force generated.

Using a bead as a handle, optical tweezers can be used to do extraordinary things with molecules. Kazuhiko Kinosita at Keio University in Japan and his co-workers attached beads to each end of an actin filament and then pulled one end hither and thither until it was threaded through a loop, creating a molecular knot (Fig. 29). They tightened the knot until it broke. Because the actin filament is somewhat stiff, like a sapling branch, it is weakened when sharply curved, and the force required to break the knotted filament was far lower than that needed to pull apart an unknotted filament.

Optical tweezers are not the only tool for handling molecules one at a time. Devices called scanning probe microscopes, devised in the

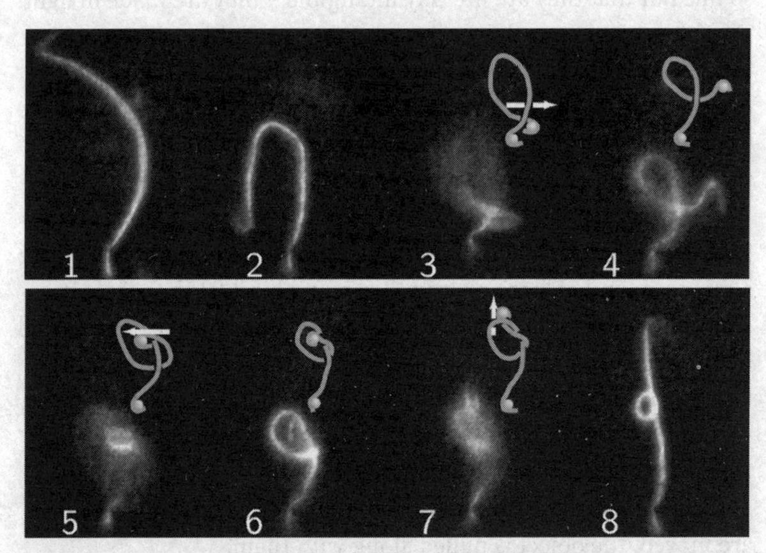

29. Optical tweezers were used to tie this knot in a strand of actin. The microscopic beads attached to each end acted as 'handles'. The actin is made visible under a microscope by shining light on it to make it fluoresce.

1980s (and used to take the images shown in Fig. 5), have proved immensely valuable not just for observing but for manipulating the molecular world. One of these instruments, called the atomic force microscope (AFM), allows researchers to probe the mechanical properties of molecules – how stiff or stretchy they are, for instance. A molecule can literally be grasped at one end by the AFM and pulled like a piece of elastic.

Motors by design

One of the most prominent prophets of nanotechnology is K. Eric Drexler, an independent scientist who heads the Foresight Institute in California. Drexler's vision, which is built around the idea of molecular-scale robotic assemblers that can put together any molecular machine (including themselves) atom by atom, has been influential on the public perception of nanotechnology's promise (and dangers), but rather less so among scientists. Some scientists worry that Drexler's idea of atom assemblers fails to take account of heat that must inevitably be released when atoms are combined. Moreover, the shapes of molecules are many and varied, but not arbitrary: there is no guarantee that a particular molecular-scale blueprint for a nanotechnological component will correspond to a stable or realizable arrangement of atoms.

Drexler first outlined his ideas in his 1986 book *Engines of Creation*, in which the protagonists (and sometimes antagonists) were nanotechnological robots. But in terms of what is already technologically feasible, a scratch-built, controllable molecular motor would in itself be an engine of creation, however primitive. With such a device, molecular-scale rods, girders, and other construction parts might be shifted into place ready for welding together.

Whereas motor proteins are powered by ATP, some researchers think that synthetic molecular motors could be light-powered. In 1999 a team of chemists led by Ben Feringa at the University of

Groningen in the Netherlands devised a molecular rotary motor, in which a rotor spins in a single direction driven by light. They exploited the process of photoisomerism: the light-induced interconversion of two different forms (isomers) of a molecule, which have the same chemical constitution but different shapes.

They constructed a molecule containing two linked propeller-like units (Fig. 30). Initially the propellers sit on opposite sides of the molecule: the so-called *trans* isomer. But ultraviolet light converts the molecule to the *cis* ('sis') isomer, in which both propellers are on the same side. So as not to bump into one another, the propeller blades twist – one upwards, one downwards. If the molecule is warmed up above 20 °C, the blades switch to the opposite configuration: that which twisted downwards now bends upwards, and vice versa. In this configuration the molecule is slightly more stable. Irradiating it with another dose of ultraviolet light then brings about the reverse switch from the *cis* to the *trans* form. But because of the propeller flip that preceded it, the *trans* form is now subtly different from that at the start: the propellers both bend down rather than up. Heating the molecule to 60 °C restores the original configuration.

The overall result of this four-step process is that one of the propeller blades makes a full revolution with respect to the other, in a predetermined direction. If the molecule is kept above 60 °C and continuously irradiated with ultraviolet light, it will spin smoothly: a light-powered molecular motor.

A different rotary device was made by Ross Kelly and co-workers at Boston College. They constructed a molecule consisting of a three-bladed propeller connected by an axle to a 'brake' that hindered the rotation of the propeller. Without the brake, the propeller rotates – but at random in either direction. The researchers aimed to use the brake to ratchet the propeller in only one direction, by executing a series of chemical reactions between

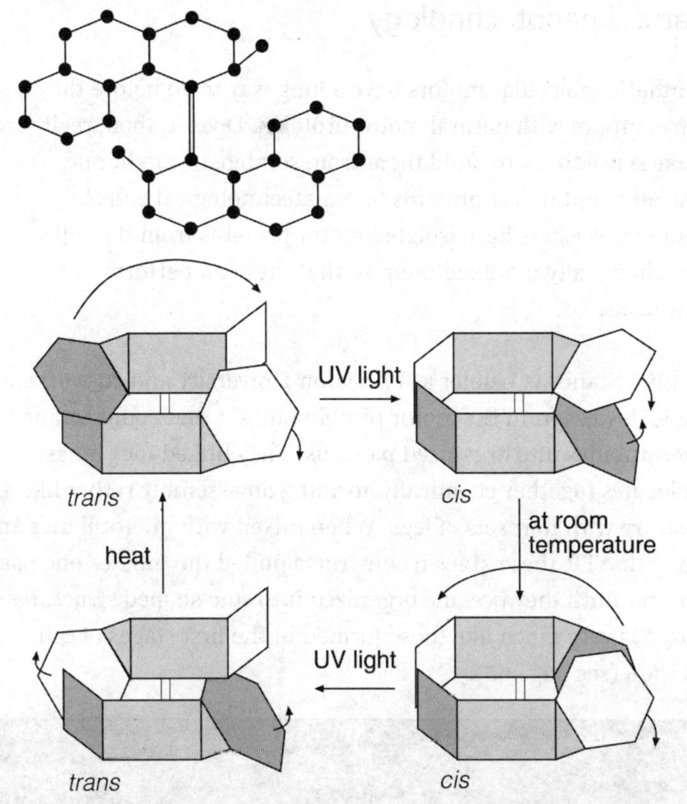

30. A scratch-built molecular rotary motor powered by light. The top image shows the carbon-atom framework.

blade and brake. But they have not yet found a way to pull their prop through more than a third of a full turn.

Both of these devices are simplistic and neither can yet be harnessed to perform a useful task. But they show how, in principle, molecular motors might be constructed. The sequence of bond making and breaking required to move Kelly's motor looks cumbersome, but a similar sequence is after all needed to produce linear motion with kinesin and myosin. At the molecular scale, such things can happen quickly enough to give the appearance of smooth motion.

Natural nanotechnology

Synthetic molecular motors have a long way to go before they can compare with natural motor proteins. Does it, then, really make sense to try to build them from scratch, or might one instead adapt motor proteins to nanotechnological ends? Some researchers have isolated motor proteins from the cell and chemically modified them so that they can perform new tasks.

In 1997 Stanislas Leibler at Princeton University and co-workers made devices from the motor protein kinesin that could arrange microtubules into organized patterns. They linked four kinesin molecules together chemically, forming an assembly rather like a creature with four sets of legs. When mixed with microtubules and fed with ATP, these kinesin constructs pulled the tubules one past another until they became organized into star-shaped structures (Fig. 31) very much like those formed in the first stages of cell division (see page 65).

31. Microtubules organized into star-shaped structures by semi-synthetic molecular motors made from modified proteins.

At the University of Washington in Seattle, Viola Vogel and co-workers have used kinesin to propel microtubules over surfaces in a selected direction. They attached kinesin molecules to a surface coated with the polymer polytetrafluoroethylene (PTFE), better known as the non-stick coating Teflon. The PTFE was applied by rubbing a block of it over the surface, whereupon the polymer film acquires striated grooves and ridges in the direction of rubbing. The polymer chains are thought to be aligned with these ridges. Kinesin molecules become attached preferentially on the ridges, which means that they form oriented rows. These rows act as linear tracks along which microtubules can be passed: the kinesin molecules pass the tubules to one another like a bucket brigade. In cells it is the kinesin molecules that are mobile and the tubules that are 'fixed'. But in these experiments the motor proteins are tethered to the surface, so their walking motions propel the microtubules instead.

The most dramatic and exciting amalgamation so far of biomolecular motors with artificial microengineering was described towards the end of 2000 by Carlo Montemagno and co-workers at Cornell University in Ithaca, New York. They commandeered a molecular *rotary* motor to turn a tiny metal propeller about 150 nanometres wide and nearly ten times as long. The enzyme ATP synthase, we saw earlier, has a head that spins on a membrane-bound spindle as it performs its task of converting ADP to ATP (see page 83). Montemagno and colleagues fixed this head to the top of a microscopic pedestal etched from nickel metal, and then they attached the metal propeller to the spindle. Under the right conditions, ATP synthase can work in reverse, breaking down ATP to ADP and spinning as it does so. The researchers initiated this process by feeding their rotors with ATP, and saw them revolve under the microscope at around five revolutions per second (Fig. 32).

Studies like this raise the exciting prospect of using molecular motors to move molecules around in a controlled way – something that brings a whole new dimension to synthesis at the molecular

a *b*

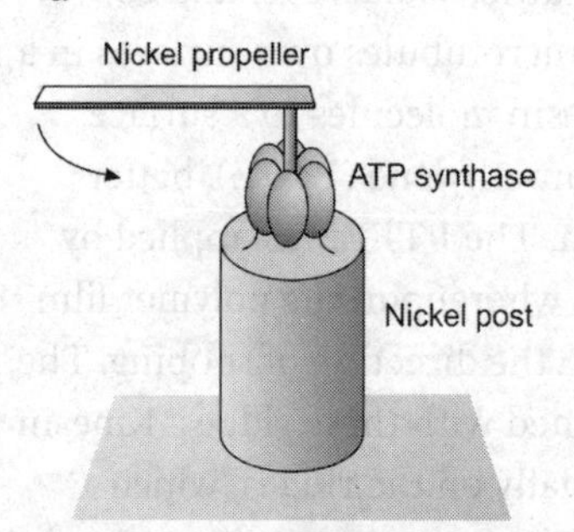

32. A microscopic metal propeller attached to the spindle of the rotary motor protein ATP synthase (*a*) rotates when the motor is driven with ATP. *b* shows an array of many such constructs. Only those for which the propellers appear here as non-vertical are working as intended: the synthesis is not yet successful in every case.

scale. No longer would chemists have to rely on the random wandering and chance encounter of molecules floating in solution: they could instead guide them precisely where they are meant to go. Because nature has already devised a wondrous array of molecular machines for such purposes, I suspect that molecular nanotechnologists will increasingly make use of the cell's machinery rather than trying to design devices from first principles. This applies not just to the generation of mechanical motion but to areas such as energy generation, sensors, and information processing. We may then see a fusion of biology with disciplines once regarded as quite different, such as mechanical and electronic engineering. Because the union will bring about results that none of the fields can achieve on its own, we could call it *biosynergetic engineering*.

Chapter 6
Delivering the message: molecular communication

Each of us is a new world. The molecular view of life reveals that the appropriate analogy for a cell is a city, teeming with molecular inhabitants. Our many-celled bodies are thus collaborations between communities. One cell communicates with another as London does with Liverpool, New York with Philadelphia: messages are passed down wires or carried from place to place by courier. Goods are transported hither and thither along the transportation network of the blood and lymph circulatory systems. It is just as Berzelius said: 'This *power to live* [is] the result of the mutual operation of the instruments and rudiments on one another.'

Molecular biology has long been content merely to document the social webs of the cell: deducing which molecules speak to which, and how they come and go. But ultimately this is not enough. We need to know also what is said, and how the messages are passed on one to another. This is information that a pharmaceutical chemist can use to develop drugs. The fundamental challenge for medical science is to learn how to take part in the body's molecular conversations: to intercept harmful or unpleasant messages, to send out new warning messages, to prevent undesirable interactions.

It is largely as a result of these efforts in the field of biochemistry that chemistry itself is undergoing something of a reinvention. The imagination of chemists is fired up by what they see to be possible in biology. Although much of the chemicals industry is devoted to the manufacture of 'passive' products – new plastics, cement, glue, paint, synthetic fibres – drug molecules were always a little different. For their task is to partake in a dynamic process, to engage in the active life of the cell. They are like actors primed for a dramatic role; indeed, they often play their part by impersonation. Now chemists are beginning to appreciate that this kind of dynamism can be achieved in purely synthetic chemical systems too. And so chemistry is becoming less about the properties of individual molecules and more about how groups of different molecules behave together – forging and breaking relationships, modifying each other's tendencies, sending out signals. Chemistry is becoming a science of *process*.

This is the attitude that underpins much of the work I discuss in this book: the development of molecular solar cells, of chemical sensors, of a molecular nanotechnology, and of molecular devices that process information. Much of the research in this area is gathered under the umbrella of *supramolecular chemistry*, which means chemistry beyond the molecule: the science of molecules in communication.

In this chapter I shall explore a few of the ways in which molecules communicate in biology, before providing a glimpse of how a similar gregariousness can be encouraged in synthetic molecules. As ever, we must remember that, while nature is inspirational, it is also parsimonious and blind. Biology uses a limited range of materials, and has a tendency to adapt a good solution endlessly for new purposes rather than exploring a completely new avenue each time. Just as the Jumbo jet is not a scaled-up pigeon, so the wise molecular engineer takes from nature principles but not blueprints.

Molecular mail

Italy and Germany became countries when their patchworks of small kingdoms and city states agreed to respect a central authority – to collaborate towards a greater good for all. The body cannot behave as a coherent entity unless cells do likewise. This means that there must be mechanisms for sending commands, edicts, and calls to action throughout the entire realm. Nerve signals from the brain are one means by which the body coordinates its actions. They are the body's telephone system.

But general messages sent *globally* throughout the body are posted like a mass mail shot into the bloodstream, in the form of molecules called *hormones*. These are diverse both in form and in function. Some hormones are large proteins; others are small organic molecules. Some are soluble in water, others insoluble (which means that courier molecules are required to carry them through the bloodstream). Some convey urgent, immediate messages such as 'Run away!' Others have long-term effects, promoting growth or the development of sexual characteristics.

All hormones are the product of the *endocrine system*, a series of glands that constitutes the overarching regulatory system for the whole body (Fig. 33). We have seen already how the hormones insulin and glucagon, produced in the pancreas, control the blood's sugar content (see page 78). Similarly, the rate of metabolic processes in cells is regulated by the hormones thyroxine and triiodothyronine released from the thyroid gland. These hormones affect energy production and oxygen consumption, in part by altering the heart-beat rate.

The control centre of the endocrine system is the hypothalamus, a gland in the brain. The hypothalamus is connected to the pituitary gland, which sits just below it, from where hormones are dispatched

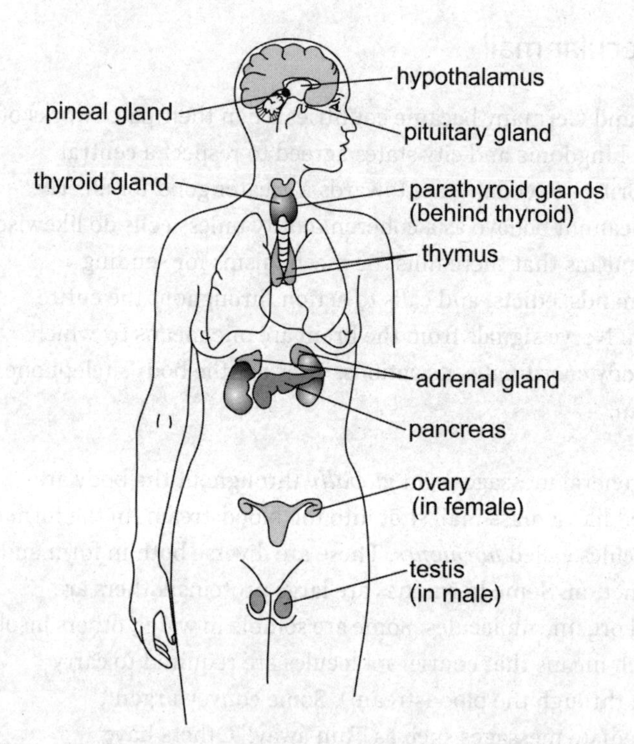

33. The body's endocrine system, a series of hormone-producing glands.

to other glands. For example, a fall in metabolic rate prompts the hypothalamus to send a molecule called thryotropin-releasing hormone to the pituitary. This gland in turn starts to send out thyroid-stimulating hormone, which triggers the thyroid gland into action.

All of the hormones released from the pituitary are peptides: small protein-like molecules. Antidiuretic hormone, for instance, controls the body's water content by regulating the production of urine in the kidneys. Growth hormone provides a stimulus for cell multiplication, and plays a key role during childhood and adolescence. It also stimulates localized growth of tissue when repairs need to be made – for example, during wound healing.

The adrenal glands manufacture some important steroid hormones. These are insoluble molecules with a carbon backbone consisting of several small rings joined together. Some steroids, such as cortisol, regulate the storage and use of the body's energy resources: the conversion of glucose to glycogen and the breakdown of proteins into amino acids. Bodybuilders and athletes use these hormones (legally or otherwise) to build up body mass and muscle.

As you might anticipate, the hormone adrenaline is also a product of the adrenal glands. Along with noradrenaline, it is released rapidly into the bloodstream in response to stress. Both hormones quicken the heart rate and dilate the blood vessels, increasing the oxygen supply to muscles so that they are prepared for extreme exertion.

The sex glands – ovaries in women, testes in men – release the hormones that differentiate the sexes and trigger changes in growth during puberty. Testosterone stimulates sperm production in men. Oestrogen and progesterone control the female menstrual cycle; their production is regulated by hormones released from the pituitary gland, called follicle stimulating hormone (FSH) and luteinizing hormone (LH).

These two hormones regulate ovulation during the menstrual cycle. In the early days of pregnancy, high levels of oestrogen and progesterone in the bloodstream inhibit the production of FSH and LH and suppress ovulation. Birth-control pills have the same effect: containing oestrogen and progesterone, they persuade the woman's body that she is already pregnant.

Production of oestrogen declines when a woman is in her thirties, and particularly during the menopause. Side effects of low oestrogen levels include an increased susceptibility to coronary heart disease and to bone loss, which are two of the prime motivations for administering oestrogen in hormone replacement therapy. The treatment remains controversial, because long-term doses of oestrogen can have unwelcome side effects of their own, including an enhanced susceptibility to breast cancer and different forms of heart disease.

Switched on

How is a hormonal message read? This depends on the nature of the message. Some hormones can be posted right through the walls of cells, wherein they bind to some receptor protein. This activates the receptor to stimulate the transcription of a particular gene, making a protein that the cell needs. This so-called direct gene mechanism of hormone action works for hormones that are small and insoluble, so that they can penetrate the fatty cell membrane.

But many hormones, particularly those comprised of peptide and protein molecules, get no further than knocking on the doors of the cell. They are received by butlers at the cell surface – receptor proteins whose job it is to convey the message to others inside the cell.

Like most other molecular communication, the passing of a message from hormone to receptor protein is an intimate affair. Molecules show no inhibitions: they speak to one another through close embraces. Lacking any other means of recognition, molecules identify one another by 'touch', through binding events in which the receptor latches onto a target (substrate) with precisely the right shape, like a key fitting in a lock. Each hormone-receptor protein on a cell surface has a binding site sculpted to fit around the hormone.

Despite the variety of messages that hormones convey, the mechanism by which the signal is passed from a receptor protein at the cell surface to the cell's interior is the same in almost all cases. It involves a sequence of molecular interactions in which molecules transform one another down a relay chain. In cell biology this is called *signal transduction*. At the same time as relaying the message, these interactions amplify the signal so that the docking of a single hormone molecule to a receptor creates a big response inside the cell.

It works like this. The receptor proteins span the entire width of the membrane; the hormone-binding site protrudes on the outer surface, while the base of the receptor emerges from the inner surface (Fig. 34). When the receptor binds its target hormone, a shape change is transmitted to the lower face of the protein, which enables it to act as an enzyme.

The process that the enzyme catalyses is the 'activation' of a so-called G protein, attached to the inner surface of the membrane. G protein is short for guanine-nucleotide-binding protein: the protein holds onto a molecule of guanosine diphosphate (GDP). When a hormone-charged receptor interacts with a GDP-laden G protein, the G protein first replaces the GDP with guanosine triphosphate (GTP, analogous to energy-rich ATP), and then breaks in two. The half that binds the GTP becomes an enzyme, and travels off to activate another enzyme at the inner surface of the cell wall. Commonly, this other is adenylate cyclase, a protein that converts ATP into cyclic AMP (cAMP).

The participants of all these processes are stuck to the cell wall. But cAMP floats freely in the cell's cytoplasm, and is able to carry the signal into the cell interior. It is called a 'second messenger', since it is the agent that relays the signal of the 'first messenger' (the hormone) into the community of the cell. Cyclic AMP becomes attached to protein molecules called protein kinases, whereupon they in turn become activated as enzymes. Most protein kinases switch other enzymes on and off by attaching phosphate groups to them – a reaction called phosphorylation. The action of a protein kinase initiates a cascade of reactions, since each activated kinase can act on several enzyme molecules, each of which in turn can do its job many times. In this way docking of one hormone to its receptor can affect many molecules inside the cell: the signal becomes amplified.

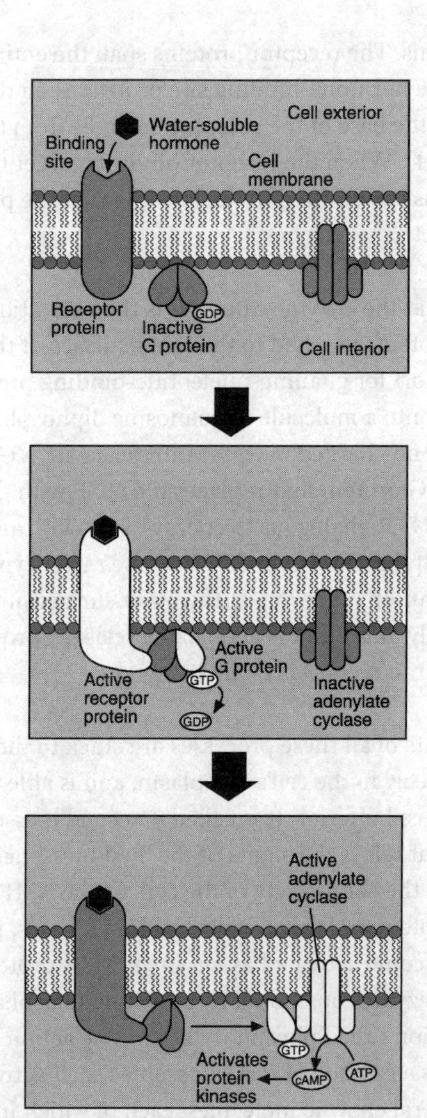

34. How G proteins work.

The process might sound rather complicated, but it is really nothing more than a molecular relay. The signal is passed from the hormone to its receptor, then to the G protein, on to an enzyme and thence to the second messenger, and further on to a protein kinase, and so forth.

The G-protein mechanism of signal transduction was discovered in the 1970s by Alfred Gilman and Martin Rodbell, for which they received the 1994 Nobel Prize for medicine. It represents one of the most widespread means of getting a message across a cell membrane. Some hormones attenuate rather than stimulate cell processes; in such cases the activated G proteins might exert an inhibitory effect on their target enzymes rather than activating them. In other cases the second messenger might be a small molecule other than cAMP: certain activated G proteins trigger the release of calcium ions from the calcium-binding protein calmodulin, for instance.

And it is not just hormonal signalling that makes use of the G-protein mechanism. Our senses of vision and smell, which also involve the transmission of signals, employ the same switching process. The roof of the nasal cavity is lined with smell sensors called olfactory hairs, which are attached to the ends of nerve cells that carry signals to the olfactory bulb – the 'smell centre' of the brain. The cell walls of the olfactory hairs are studded with receptor proteins designed to bind particular odorant molecules that enter the nose.

There are several hundred different kinds of odorant receptors, each one with a binding site shaped to accommodate a specific common odorant. We can, however, discriminate between a wider range of odours than this, because each odour is typically the result of a complex blend of different odorant molecules. The olfactory bulb forms an 'image' of the smell from the mixture of impulses that it receives from different receptors, much as we recognize a person's face from the sum of the different component parts.

In smell signalling, the cAMP produced by G proteins binds to a membrane protein called a sodium channel, whereupon the channel opens up and lets sodium ions flow into the cell. This triggers a nerve impulse, which passes to the olfactory bulb. The same basic process generates visual signals in the optic nerve when stimulated by light.

Our sense of taste is due largely to our olfactory system. The taste buds in our tongues can distinguish only relatively crude signifiers of taste: sweetness, bitterness, saltiness, and sourness. The full delight of a matured cheese or freshly baked bread comes mostly from the odorant molecules they release.

All in the mind

Hormones can trigger complex webs of biochemical action, but the messages they bear are pretty crude, related to the exigencies of growth and survival. It is quite another matter that communication between molecules gave birth to the Sistine Chapel, to *The Magic Flute*, to the theory of relativity. Yet the mind is, after all, made of molecules.

At the same time, the mind is still a mystery – one of the great remaining mysteries of science. Some scientists argue that the mind will never be able fully to comprehend itself, that the self-referential nature of the problem will always create blind spots. Others believe that a scientific explanation of consciousness is on the horizon. Either way, it is likely that the secrets of the mind lie far beyond the molecular realm, embedded in questions about the behaviour of complex, highly connected information networks. Here we see the limitations of reductionism – for the molecular processes of thought are now fairly well mapped out, yet their collective consequences are barely sketched.

The brain contains somewhere in the region of a billion to a hundred billion brain cells or *neurons*. That is nothing to write home about – other organs are comparably populous. But the distinguishing feature of the brain is the complexity of the communication network between these cells. Each neuron makes around a thousand links, so there may be up to a hundred trillion interconnections in the brain – about the same as the number of stars in a thousand galaxies like our own. On such a transportation network, you would be lost in an instant. The degree of connectivity in the integrated circuits of computers is nowhere near so great, and it is no surprise that computers, for all their literal-minded speed, fail miserably at some tasks that a child can do in a flash.

Neurons send nerve signals – in essence, electrical pulses – to one another along tubular channels called axons. The axon ends in a series of branches whose tips push up against the membranes of other neurons. At these junctions, called synapses, a nerve signal is transmitted from one neuron to another. Neurons also sprout many shorter, bushy branches called dendrites, which collect information from the axons of other cells. The axons are, if you like, the motorways of the brain, stretching from one neuronal city to the next. They end in slip roads that connect up at synapses to the city road system of the dendrites.

Although axon signals are electrical, they differ from those in the metal wires of electronic circuitry. The axon is basically a tubular cell membrane decorated along its length with channels that let sodium and potassium ions in and out. Some of these ion channels are permanently open; others are 'gated', opening or closing in response to electrical signals. And some are not really channels at all but pumps, which actively transport sodium ions out of the cell and potassium ions in. These sodium-potassium pumps can move ions 'uphill' – from regions of low to high concentration – because they are powered by ATP.

In its 'resting' state, the axon has an imbalance of sodium and potassium ions inside and outside that sets up a charge difference, or voltage, across the membrane: the fluid inside has a small

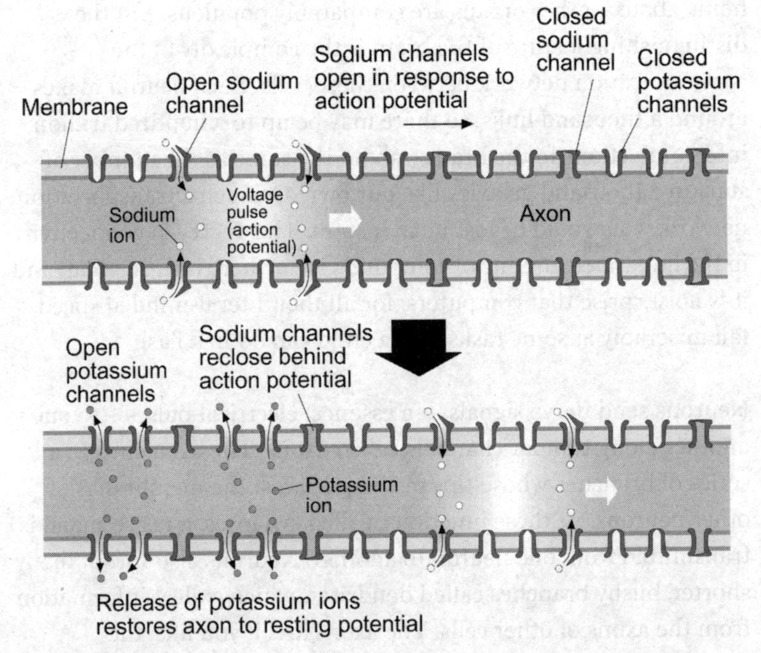

35. Electrical pulses are sent down the axon by the opening and closing of ion channels.

negative charge (the 'resting potential') relative to that outside. When a signal is sent down the axon, some of the gated sodium channels open up, altering the distribution of ions and reversing the imbalance: the inside becomes positively charged relative to the outside. This region of reversed voltage opens up sodium channels ahead of it, so that it moves along the axon. At the same time, the channels behind close up and the resting potential is restored. In this way, a voltage pulse or 'action potential' travels down the axon (Fig. 35).

At a synapse, this nerve impulse is transmitted from the axon to another neuron. The signal is generally first converted from electrical to chemical form. A small molecular messenger called a neurotransmitter conveys the signal across the space (called the synaptic cleft) between the terminal membrane of the axon and the membrane of the other neuron. The neurotransmitter is packaged up inside a bubble-like membrane that merges with the axon's cell wall, releasing the molecular message into the synaptic cleft. It travels to the outer surface of the other neuronal membrane, where it becomes bound to receptor proteins.

A diverse array of molecules serve as neurotransmitters. Some are simple amino acids – glycine and glutamate – or molecules derived from them, such as serotonin and dopamine. The molecule acetylcholine is a neurotransmitter that carries messages between the central nervous system and muscle cells at neuromuscular junctions (see page 103). When acetylcholine binds to its receptor on a muscle cell, the receptor is transformed to an open sodium channel. Sodium ions rush into the cell, changing the voltage across the cell membrane and opening up a voltage-controlled calcium channel. This triggers a rise in calcium ion concentration inside the cell, which stimulates muscle contraction.

Acetylcholine illustrates the general function of a neurotransmitter: to open or close an ion channel, thereby altering the voltage across the membrane in which the receptor sits. This converts a chemical message back to an electrical one. Acetylcholine does the job directly, since its receptor is itself an ion channel. Some other neurotransmission pathways function rather differently: they use a second messenger to transfer the message from the neurotransmitter to an ion channel, again with the mediation of G proteins.

Is it surprising that the G-protein signal transduction mechanism appears in so many different contexts? Not really. As the complexity of multicelled organisms evolved, cells with ever more specialized functions arose from common ancestors with more general functions. Tried-and-tested mechanisms for certain tasks would be retained, though adapted where necessary. That, after all, is why we share genes with yeast and bacteria. The G-protein pathway is an effective way of passing a chemical message from the outside to the inside of a membrane, and amplifying it in the process. The cell's motto is: if it works, find a way to use it.

Pleasure and pain

Neurotransmission is a common target for drugs – beneficial, harmful, pleasurable, or, depending on the dose, all three. The nervous system is one of the most vulnerable parts of the body: if nerve impulses are blocked, we cannot move. Many animals make toxins that cause paralysis in their prey by attacking the sodium-potassium pumps or the voltage-gated ion channels in axons, blocking the progress of action potentials.

Muscle action is also affected by drug molecules that resemble acetylcholine and so compete with it in binding to the receptor proteins at neuromuscular junctions. Nicotine, the active ingredient of tobacco, is one such: it binds to a certain class of acetylcholine receptors in muscle and causes the associated stimulatory sensations: increase in heart rate and dilated pupils. Why the sensation is pleasurable is not, however, fully understood. Curare is a lethal toxin present in the bark of a South American plant, which was once extracted and used by the indigenous people to poison arrow tips. Curare binds to the same class of acetylcholine receptors as nicotine, but does not activate them – so muscle action is prevented. An animal poisoned with curare will die of asphyxiation, unable to inflate its lungs. The medieval poison hemlock works in the same manner.

Whereas some neurotransmitters stimulate neurons, the role of others is to quieten them: to suppress the firing of action potentials. These are said to be inhibitory, and include glycine and the molecule gamma-aminobutyric acid (GABA). Our thoughts are a complex interplay of stimulation and inhibition, as neurons weigh up the various signals they receive from their neighbours and decide whether or not, on balance, they should fire off a salvo themselves.

Hallucinogenic drugs such as LSD (lysergic acid diethylamide) and mescaline overexcite the brain by enhancing the stimulatory effects of serotonin. The poison strychnine blocks inhibitory signals, leading to uncontrollable muscle spasms and a particularly unpleasant death. Depressants assist the binding of inhibitory neurotransmitters or (like alcohol) interfere with the action of excitory neurotransmitters.

Drugs that relieve pain typically engage with inhibitory receptors. Morphine, the main active ingredient of opium, binds to so-called opioid receptors in the spinal cord, which inhibit the transmission of pain signals to the brain. There are also opioid receptors in the brain itself, which is why morphine and related opiate drugs have a mental as well as a somatic effect. These receptors in the brain are the binding sites of peptide molecules called endorphins, which the brain produces in response to pain. Some of these are themselves extremely powerful painkillers.

Cannabinoids, the active ingredients of cannabis, also bind to inhibitory neuroreceptors in the brain to produce pain relief. The natural target of these receptors is a molecule called anandamide, which, like endorphins, is produced in response to pain signals. A closely related molecule called oleamide seems to be the biochemical trigger that induces natural sleep.

Not all pain-relieving drugs (analgesics) work by blocking the pain signal. Some prevent the signal from ever being sent. Pain signals are initiated by peptides called prostaglandins, which are manufactured and released by distressed cells. Aspirin (acetylsalicylic acid) latches onto and inhibits one of the enzymes responsible for prostaglandin synthesis, cutting off the cry of pain at its source. Unfortunately, prostaglandins are also responsible for making the mucus that protects the stomach lining (see page 78), so one of the side effects of aspirin is the risk of ulcer formation.

One of the surprising recent discoveries in neuroscience was that extremely small inorganic molecules can also act as neurotransmitters. Carbon monoxide and nitric oxide – both of them two-atom molecules – serve this function. They are both poisonous in large doses, because they compete with oxygen in binding to haemoglobin. But 'the poison is in the dose', and in small amounts nitric oxide does some important things. It triggers the dilation of blood vessels, which can relieve stress on the heart. This is why nitroglycerin, which decomposes to release nitric oxide, is administered to treat heart problems. The improvement in circulation initiated by nitric oxide provides the basis for the drug Viagra, which is used to treat erectile dysfunction in men.

Supramolecular chemistry

In recent decades, scientists have become interested in mimicking, in synthetic systems, some of the molecular communication processes of the cell. There are many motivations for this. Drug development is often a matter of concocting a good disguise, so that a synthetic molecule will pass itself off as a natural one and bind preferentially to a receptor, blocking or initiating a biochemical signal. Signal transduction in the eye's retinal cells and in the olfactory system suggests the concept of molecular sensors that can detect light or other molecules with high sensitivity. Molecular engineers are looking to the olfactory apparatus for inspiration in designing 'artificial noses' that can identify complex mixtures of molecular components.

The principle loudly extolled by nature is the 'lock and key': molecules get together when one fits with the other.* To turn such a 'recognition' event into a communication process, the binding event should trigger some change in the receptor that allows it to relay the signal downstream. In biology this relay process is commonly catalytic: binding turns the receptor into an active enzyme. But the

* This metaphor was first used by the German chemist Emil Fischer in 1894 to explain how enzymes are so selective about the transformations they catalyse.

signal might also be passed on in other ways: by the emission of light or release of an electron, for instance, or (as in the case of the acetylcholine receptor) by the creation of an electrochemical potential.

Building artificial signal-transduction processes at the molecular scale is a common objective in supramolecular chemistry. From its outset, this discipline was biologically inspired. In the 1960s, the French chemist Jean-Marie Lehn investigated so-called crown ether molecules that would recognize and bind specific metal ions. Lehn was interested in transporting ions such as sodium and potassium across lipid membranes. Although this can be mediated by protein channels and pumps, another strategy is to engulf the ion within a molecule that will 'dissolve' in the fatty interior of the membrane wall. Such molecules exist in nature, and are called ionophores. A typical example is valinomycin, a ring-shaped peptide with a central hole into which a potassium ion will fit. Crown ethers are synthetic mimics of valinomycin: they too are ring-shaped, and will bind a metal in their central cavity. The metal ion is held more or less securely depending on the relative sizes of the ion and the hole. If the hole is too big, the metal 'rattles around' and is only loosely bound; too small, and it will not fit. So crown ethers can be tailored to fit specific metal ions – to display molecular recognition, in other words.

By the 1970s, Lehn and others were making synthetic receptor molecules of all shapes and sizes, with cavities designed to accommodate a wide range of inorganic and organic targets. These 'guest' molecules are held in place by interactions with their hosts that are weak relative to the covalent bonds that hold the atoms together in the molecules themselves. In this way the guests can be picked up and released again. That is how valinomycin works as a metal-ion transporter: it captures the ions on one side of the membrane and lets them go again at the other side. Supramolecular chemistry is essentially about bringing molecules together into loose associations that can be disassembled back into their separate components.

When crown ethers take up a metal ion, they change shape. Alone, they are rather loose, floppy rings, like rubber bands. With a metal ion in their core, they become organized into relatively rigid structures in which the ring contains zigzag-like kinks: a crown, in other words (Fig. 36). Shape changes of this sort are common when a receptor binds its target.

If binding alone is the objective, a big shape change is not terribly desirable, since the internal rearrangements of the receptor make heavy weather of the binding event and may make it harder to achieve. This is why many supramolecular hosts are designed so that they are 'pre-organized' to receive their guests, minimizing the shape change caused by binding.

But if the idea is to use binding as the trigger for relaying some signal, then a shape change is often crucial. A particularly dramatic change in shape induced by host–guest binding was reported in 2000 by Ulrich Koert and colleagues at the Humboldt University in

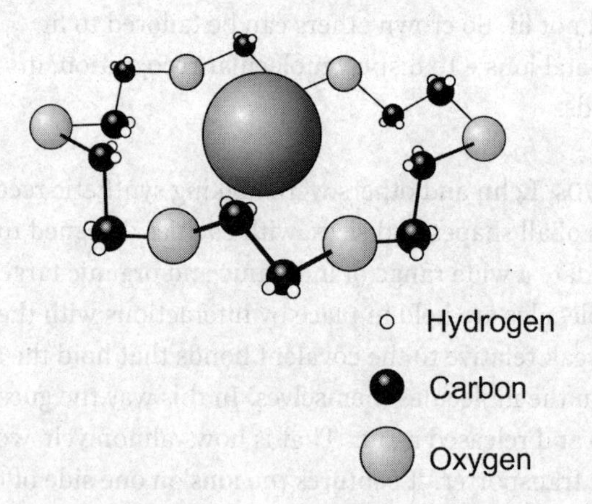

36. A crown ether is a cyclic molecule that captures a metal ion in its central cavity.

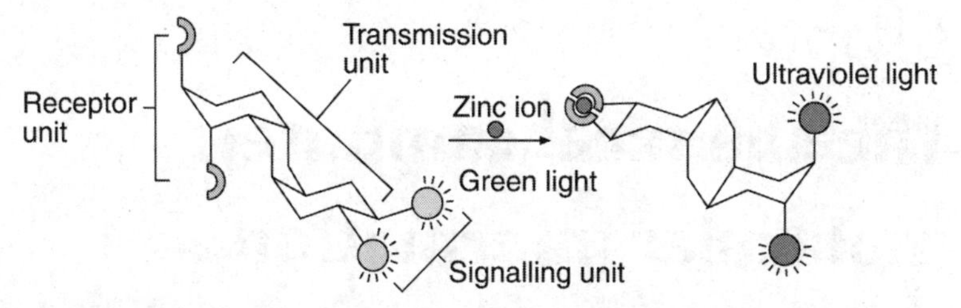

37. An artificial transducer molecule that converts binding of a zinc ion into a signalling event, by changing shape and altering its fluorescence properties.

Berlin. They constructed a receptor molecule that can be considered as a series of 'modules': two arms, two legs, and a 'transducer' unit like a flexible torso connecting the arms to the legs. When the two arms close around a zinc ion, the transducer unit flips and pulls the two legs far apart (Fig. 37). The legs are tipped with fluorescent groups, which change their emission wavelength – from green to ultraviolet – when the distance between them increases. The researchers pointed out how this molecular receptor shows some of the features of a protein receptor involved in signal transduction, responding to binding of the target at one end by altering its shape, and thus its behaviour, at the other end.

A shape change that alters a molecule's light-emitting properties has been engineered in several other synthetic receptors. But using recognition and binding to switch a molecule's *catalytic* behaviour, as in the G-protein signalling mechanism, is rather more challenging, since this means ensuring that the final shape is exactly what is needed for the catalyst to do its job. Harder still would be the task of organizing several molecules into a relay that would carry a message downstream. All the same, the skill of supramolecular chemists is increasing daily, and it would not be at all surprising if we were soon to see artificially devised molecular communication systems that approach the sophistication of those the body uses to run its realm harmoniously.

Chapter 7
The chemical computer: molecular information

In the end, we are left with the same question: what is life? It will not be answered – or at least, not here. But Schrödinger's answer – negative entropy (see page 74) – for all its shortcomings contains a grain of truth. For it is a necessary but not sufficient characteristic of life that it imposes order on chaos. Chaos is death. If cells cannot send and receive clear messages, if they do things at the wrong time, if their membranes lose their organization, if proteins fail to fold, then life cannot be sustained. We are islands of order in a wild world.

Where does this order come from? Organization can arise spontaneously in inorganic matter too: think of the serried ranks of mare's-tail clouds, or the regular ripples in wind-blown sand. It seems highly likely that this kind of 'self-organization', which can arise unbidden in systems fed with energy that prevents them from achieving a static equilibrium, has a part to play in life's orderliness. But that is not enough. The kind of coordination needed for a cell to copy its chromosomes and divide in two, to make a functioning and reproducible protein molecule, or indeed to grow from a single fertilized egg into a multicelled Mozart, cannot rely on the 'blind' patterning processes that paint the skies and the deserts. There needs to be a firmer hand on the wheel.

We have all heard of what this guiding hand consists. It is DNA, a string of molecular beads chopped and bundled into our forty-six little X-shaped chromosomes. The human genome – our full complement of DNA – is often called the 'book of life'. As I write, scientists have just finished decoding the first draft of this book: they have sketched out in broad detail the molecular messages in each of the chromosomes.

Incautious things are said about the project to map the human genome. One hears, for instance, that a sufficiently skilful engineer could make a human from the information therein. This is nonsense. The body is full of molecules that are not encoded in the genome – it encrypts only proteins, and even those in somewhat garbled and incomplete form. The genome tells us nothing about the lipids that make up cell membranes, let alone about how they are driven by physical forces to aggregate into sheets, loops, and spheres. The genes will not tell us how neural signalling works, how the brain encodes thoughts and sensations in delicately timed trains of electrical pulses. There is no gene for bone, for tooth enamel. The genome is the book of the cell in much the same way as the dictionary is the book of a performance of *Waiting for Godot*. It is all in there, but you will not deduce one from the other.

Nevertheless, the genome *is* an instruction booklet in molecular form. It tells us how to make proteins, the molecules that orchestrate the spectacular molecular performance of life. In this chapter I want to say a little more about the nature of that script: how it is read and enacted. But my ultimate intention is broader. For the molecular scientist, genetics says something truly dramatic and profound about molecules: they can carry and transmit information. The theoretical biologist John Hopfield of Princeton University points out that this is one of the many ways in which biology provides an 'existence theorem' to inspire chemists. 'Mathematicians use this term', he says, 'in reference to a proof that some function which they want to construct actually does exist, and is not impossible. In this sense, the observation of birds flying provides an existence theorem that an engineer should be able to design a flying machine.'

In the same way, genetics shows that it is possible to perform computing with molecules. For computing is nothing but the storage, transmission, and processing of information, all of which the genetic machinery can do. Says Jean-Marie Lehn, 'there is a "molecular logic of living organisms"'.

This is really a corollary of the previous chapter, where we saw that molecules can communicate with one another. A genuine molecular logic is more precise: it requires not simply that one molecule affects the behaviour of another, but that they can transfer and manipulate encoded information in well-defined ways. This is how a computer works, by passing data between switches and memory devices made from semiconducting and magnetic materials.

Computing with molecules is just one of the ways in which the dimension of information is entering molecular science. More generally, chemists are becoming accustomed to the idea that molecules can be programmed to behave in certain ways: that their properties can be written into the fabric of the molecule just as a set of instructions can be programmed into a robot. Lehn says 'The outlook . . . is toward a general science of informed matter.' Such a chemistry is a genuinely new science, in many ways quite different from the traditional chemistry that makes useful substances. It is a science that is about an active 'becoming' rather than a passive 'being'. It is already happening, and we do not know where it might take us.

How the cell 'becomes'

Every book is written in a particular language, and the genome is no different. The language of the genes is a simple code, whose characters are the four nucleotide molecules that represent the beads of DNA's molecular necklace (see page 42). Each of these molecules contains a so-called base, which encodes the information. There are four DNA bases: adenine, cytosine, guanine, and thymine (A, C, G, and T). As DNA is a linear polymer of nucleotide units, the information it uncodes can be represented as a linear string of these four characters. Part of it might look something like this:

GTGGATTGACATGATAGAAGCACTCTACTATATTC

A four-letter alphabet might seem a rather limited system for writing complex messages. But, if we think of this sequence as a code rather than strictly as an alphabet, it can be as complex as you like. We could, for example, denote every letter in the Roman alphabet by a series of several bases: GTG for 'a', GAT for 'b', and so on. The number of permutations of four characters in groups of three is sixty-four – more than enough to encode all the alphabet. Using such a code, we could write the Bible as a string of A's, G's, C's, and T's.

The cell does not have much use for the message of the Bible; what it needs are messages for making proteins. The way that a protein chain folds up is determined by its amino-acid sequence (see page 41) – so the 'information' for making a protein is uniquely specified by this sequence. DNA encodes this information using a cipher just like that suggested above: groups of three bases represent each amino acid. This is the genetic code.*

* I should point out that here I have made one of the many necessary simplifications in descriptions of genetics. We have seen that some of the most important parts of proteins, such as the haeme unit of haemoglobin, do not consist of amino acids but of other chemical groups. These so-called prosthetic groups are added to the protein chain after it has been manufactured, and are constructed by other enzymes. The bare protein chain, without prosthetic groups, is called an apoprotein. Generally it is completely useless until the frills have been added. Truly to deduce the structure and form of a protein, we need to know not only its amino-acid sequence – and thus the sequence of the gene that encodes it – but also the identity and function of proteins that operate postnatally, so to speak, on the apoprotein.

How a particular protein sequence determines the way its chain folds is not yet fully understood. This means that we cannot deduce a gene's function from its sequence alone (although we can sometimes make good guesses). The first draft of the human genome is full of genes of unknown purpose.

Nevertheless, the principle of information flow in the cell is clear. DNA is a manual of information about proteins. We can think of each chromosome as a separate chapter, each gene as a word in that chapter (they are very long words!), and each sequential group of three bases in the gene as a character in the word. Proteins are translations of the words into another language, whose characters are amino acids. In general, only when the genetic language is translated can we understand what it means.

DNA is a double-stranded polymer: two chains are twisted around one another in a double helix. Each of these strands consists of a string of nucleotides, embodying encoded information. But the strands are not identical. They are stuck together by a kind of zip of hydrogen bonds (see page 41) between the bases on one strand and those on the other. Although all the bases can form hydrogen bonds, they have distinct preferences in their pairing relationships: A sticks to T, and G to C. So the twin helices of DNA have complementary sequences: wherever an A appears in one strand, a T appears in the other, and so forth. This means that each gene is written in two versions, echoing each other in a kind of mirror language.

The particular partnerships of bases are dictated by their shapes. The A and G bases are similar molecules, as are C and T. So an A–T pairing has much the same overall shape and size as a G–C pairing. These pairs point inwards between the two coiled strands like rungs on a spiral staircase. Because the rungs are the same size, the strands coil evenly. An A paired with a G would create a bulge, and the resulting distortion would destabilize the pairing. Likewise, if

C and T were to pair up, there would be a constriction in the double helix. A–C and G–T pairings, meanwhile, are discouraged by the disposition of hydrogen-bonding groups in these base pairs. So the pairing preferences are a matter of a good fit – of *complementarity* – between the two partners.

This is one of the crucial points about biological information flow: the transmission of data occurs via molecular-recognition processes, which ensure that each part of the message is read correctly.

DNA is replicated – the genome is copied – when a cell divides. Because the two strands are complementary, each of them can serve as a template on which the other can be put together. If A always binds preferentially to T and so forth, the sequence of a 'naked' single strand will guide unlinked nucleotides to line up in the right order for forming the complementary strand.

In order to act as a template, a single strand is unzipped from its partner by special enzymes. Complementary strands are then pieced together along both of the exposed strands; an enzyme called DNA polymerase catalyses the addition of each new nucleotide. So the two new double helices each contain one strand from the original.

Although enzymes help the process along, the essential information needed for the copying process is already written into the DNA templates. In the early 1980s, Leslie Orgel at the Salk Institute in California and co-workers showed that individual nucleotides can be assembled into polymers on a template of complementary nucelotides without assistance from enzymes. A string of eight C-bearing RNA nucleotides, for example, will act as a template for joining together eight G-bearing nucleotides. Orgel had to cheat a little, however, by using G nucleotides that had been 'activated' by the addition of a reactive chemical group, which helped them join together.

This template-assisted polymerization is not, in itself, replication: the new strand is complementary to the template, not identical. The first example of genuine artificial molecular replication was reported in 1986 by German chemist Günter von Kiedrowski. He used the same templating process, but chose a template that was self-complementary: it was its own complement. The template was a six-nucleotide DNA molecule with the sequence CCGCGG. Its complementary sequence is the same as this, because the two strands of the helix line up with one oriented in the reverse direction to the other. Von Kiedrowski assembled this complement from two three-nucleotide fragments, again activated to help them link up (Fig. 38).

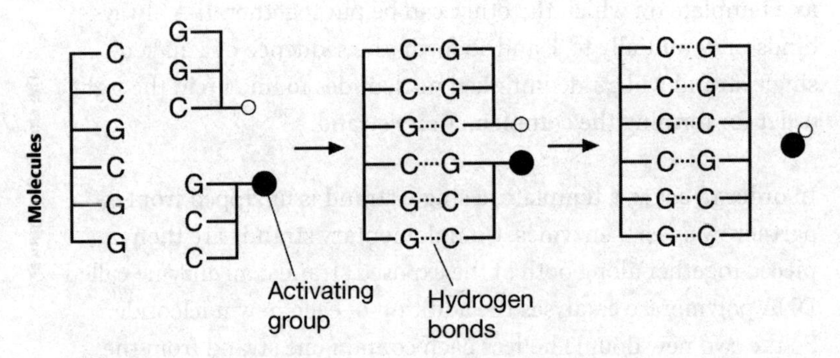

38. Molecular replication in a six-unit nucleic acid.

Errors and junk

At some stage during the production of this book I will have received from the publishers page proofs – rough versions of the final pages, prepared from the manuscript I provided. It will (I hope) be a more or less faithful transcription of what I have written. But almost without doubt there will be a scattering of small errors owing to typing mistakes or file-reading glitches. No writer is surprised by such things, because it is inevitable that copying a long and complex message introduces a few mistakes.

The same is true in the genetic processes of transcription (where DNA is copied into RNA) and translation (where RNA is copied into a protein sequence). Occasionally a wrong nucleotide or amino acid will be inserted into the chain, for molecules are not capable of perfect recognition all the time. Probably about one in every twenty or so proteins is incorrectly made.

Does this matter? On the whole, no. It will be an unusual situation if, between me and the publishers, we spot all the typographical errors in this book before it goes to print. But hopefully none of them will be so serious that you are unable to catch my meaning. Similarly, in proteins most of the chain is scaffolding, which holds in place the few amino-acid residues that conduct the protein's catalytic task. An error here and there in the scaffolding might not be serious. Sometimes the consequence of an error might be a completely non-functional molecule; but because the cell makes not one but typically dozens or hundreds of enzyme molecules for any particular task, one or two duds do not matter.

Here I am talking about *random* errors. Far more serious are *systematic* errors, which arise further back in the stream of biological information flow – closer to the ultimate repository of information. A wrongly transcribed RNA molecule could generate hundreds of faulty proteins. So there are enzymes that check the transcription process quite carefully for copying errors, reducing their frequency to around one in ten thousand.

Even errors in transcription are seldom a matter of serious consequence: after all, RNA molecules are ephemeral; the cell can always make more of them. But an error in DNA can be bad news indeed, because there is no way of correcting it once it is in place. One misplaced nucleotide in a gene means that all the RNAs made from that gene, and all the proteins made from those RNAs, contain the analogous fault. Worse still, every cell stemming from division of the genetically faulty cell inherits the same flaw. If the genetic defect occurs in a gamete – a sperm or egg cell – then it is transferred to progeny derived from that gamete. This is why DNA replication is scrutinized extremely carefully by 'proof-reading' enzymes, which permit no more than one error in a billion bases to creep through. Without these molecular proof-readers, we would acquire about 1,000 defective genes in every new cell.

Inheritable errors, the result of mistakes made in DNA replication during production of the gametes, are known as mutations. Once established, they are passed on from parent to offspring right down the genealogical tree. Mutations are responsible for genetically related disorders such as cystic fibrosis, as well as for genetically linked predispositions to conditions such as cancer and heart disease. But, in spite of such terrible consequences, mutations are also the spice of life. Indeed, we owe to them our very existence. If the primitive single-celled organisms replicating in the steamy broth of the early Earth had not picked up occasional mutations – if they had invariably copied their DNA without a single mistake – there would have been no evolution, no emergence of complex life.

Something else will doubtless have happened to my text when the proofs come back from the publishers. Every so often there will be words I did not write. They will not be errors, however, but will make perfect sense: they will be changes made by the editor, and will, I am sure, make the text far easier to read and understand than was my original.

It came as something of a surprise in the mid-1970s to find that genes too need editing. The RNA transcript that peels off from the DNA template is not fit for translation to proteins, for it contains a lot of useless information. These 'primary RNA transcripts' are rather like sentences that have fragments of other sentences inserted seemingly at random. The RNA molecules need heavy editing before they present a clear message fit for translation.

The useless inserts are called introns, and sometimes they make up most of a gene. They are also known as non-coding sequences, since they do not encode parts of proteins. Enzymes snip out the introns from the RNA primary transcript, and splice together the two ends of the coding regions (called exons).

This is just one of the ways in which the supposed 'book of the cell' is mostly garbled nonsense – or boring repetition. It is thought that only about 2–3 per cent of the entire human genome codes for proteins. Some sequences get repeated for good reason. Each human chromosome ends in the sequence TTAGGG repeated about 2,500 times. These sections, called telomeres, are thought to keep the chromosomes stable. They get shortened each time a cell divides, and their eventual erosion contributes to the ageing process. But many other repeat sequences serve no useful function. Transposons are repeats that jump about in the genome, leaving copies as they go. They are thought to be a genetic parasite living in the very core of our being, whose only purpose is to replicate themselves. Introns may be the remnants of ancient transposons that lost the ability to move on.

The protein machinery for cutting, splicing, replicating, and synthesizing nucleic acids provides the principal tools for genetic biotechnology, the manipulation of genomes. Restriction enzymes, for instance, are proteins that can recognize a specific short sequence of DNA and cut the chain at that point. Ligases join loose ends of DNA together. Stretches of DNA can be replicated indefinitely in a test tube using DNA polymerases. The double strands are separated by heating, exposing them for templated replication. Repeated cycles of replication and heating multiply the DNA exponentially. This process, called the polymerase chain reaction (PCR), uses a DNA polymerase taken from a bacterium that lives in hot springs. The enzymes of this bacterium have evolved to withstand high temperatures, and so this DNA polymerase is not destroyed by the heating cycles.

These tools allow scientists to 'rewrite the book': to insert new genes into an organism's genome. Crop scientists are interested in developing plants with genes that confer resistance to pests, or to drought, or to particular herbicides, as well as incorporating genes that improve the flavour of the crop, or its growth rate, or whatever. One potential danger is that genes for herbicide resistance, for example, might become transferred from agricultural crops into weeds, generating new breeds of 'superweed'. The likelihood of this 'horizontal' trans-species transfer of genes is not known.

Some people object to genetic engineering on the grounds that it is ethically wrong to tamper with the fundamental material of life – DNA – whether it is in bacteria, humans, tomatoes, or sheep. One can understand such objections, and it would be arrogant to dismiss them as unscientific. Nevertheless, they do sit uneasily with what we now know about the molecular basis of life. The idea that our genetic make-up is sacrosanct looks hard to sustain once we appreciate how contingent, not to say arbitrary, that make-up is. Our genomes are mostly parasite-riddled junk, full of the detritus of over three billion years of evolution. There seems little that is admirable or elegant in this unruly library; rather, the admiration should be reserved for the cohorts of diligent proteins that painstakingly sift snippets of meaning from reams of nonsense. It is truly amazing how well the whole affair works; but, like most of life, it is a makeshift compromise in which efficiency and tidiness count for little.

Construction blueprints

DNA is a supreme example of 'informed matter'. It is programmed to assemble in a highly specific manner, each nucleotide marrying up with its complement over thousands of base pairs. This kind of programmed self-assembly represents one of the goals of supramolecular chemistry. Atoms do not in themselves display many powers of discrimination; but by making the molecule, rather than the atom, the fundamental building block, supramolecular chemists are able to programme much more guiding information into their bricks and mortar.

Yet DNA provides more than an existence theorem for programmed self-assembly. It can supply the very fabric. Why not use the principles of complementary base pairing to join together DNA girders into structures much more complex than the double helices of the cell?

This concept has been explored by Nadrian Seeman at New York University. He commandeers the enzymatic apparatus of biotechnology to cut and splice DNA into remarkable edifices, such as cagelike polyhedra: a cube and a truncated octahedron (Fig. 39). The edges of these structures are DNA double helices, but at the corners three coils meet in a triple junction. These junctions are cunningly woven: the twin strands go separate ways along different edges, where they intertwine with new strands. Seeman and his co-workers make these junctions by piecing together synthetic DNA with carefully planned sequences.

The trick is then to assemble triple junctions into a three-dimensional geometric molecular object. Seeman gives the branches 'sticky ends' where one strand extends beyond the other.

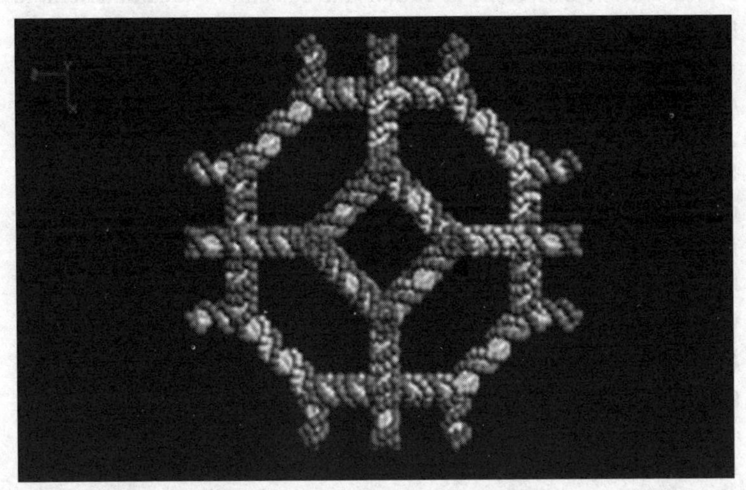

39. A polyhedral molecular assembly made from double-stranded DNA.

Here the exposed, unpaired bases are ready to match up with those on another strand – but only if it has a complementary sequence. In this way, the ends are selectively sticky and can be assigned one to another, so that the whole structure is programmed to build itself from its component parts. Once the sticky ends are married up to their partners by base-to-base hydrogen bonding, ligation enzymes forge strong bonds to secure the backbone.

For the present time, these molecular constructions are follies of virtuosity: demonstrations of the astonishing control that molecular recognition can provide over nanometre-scale architecture. But Seeman suggests that his DNA frameworks might act as scaffolds for assembling other molecules and materials in useful ways. For example, it is possible to coat DNA strands with silver, transforming them into electrically conducting molecular wires. Might we one day build tiny electronic circuits by 'genetically' programming the wires to link up in a specified pattern?

Moreover, sticky-ended DNA can clip together molecular-scale objects in a selective manner. Chad Mirkin and colleagues at Northwestern University in Illinois have used this idea to assemble little particles of gold into clusters. The particles are just a few nanometres in size – so-called nanocrystals. Each has a tag consisting of single-stranded DNA, but, because the tag sequences are not complementary, the particles remain separate. By adding lone DNA strands whose two ends complement the sequences of the tags, the researchers are able to link the nanoparticles together. The resulting clusters scatter blue light strongly, so the solution turns the colour of wine. Mirkin and colleagues are now developing this technique commercially as a method for simple visual detection of DNA strands with a particular sequence – something that is commonly required in genetic analysis.

By assembling nanocrystals of metals or semiconductors into organized arrays, some researchers hope to be able to build electronic devices far smaller than those currently made with the conventional microfabrication techniques used for creating silicon chips. Semiconductor nanocrystals could act as memory elements for storing electronic information, and they interact with light in ways that could be useful for making light-based information-processing devices. Programming the assembly of nanocrystals using DNA linkers might provide one way of arranging them into circuit patterns. Another possibility has been suggested by the work of Angela Belcher of the University of Texas and co-workers, who used the molecular-recognition properties of proteins rather than DNA. They developed small peptide molecules that would recognize and stick to the surfaces of different kinds of semiconductor. The peptides could 'feel' how the atoms were organized at the surface of the semiconductor crystals. Genetically engineered motor proteins (see Chapter 5) with peptide arms that recognize certain kinds of semiconductor might one day be used to drag nanocrystals around on a molecular building site and arrange them in a circuit pattern.

Molecular logic

Since the invention of the computer in the 1940s, the computing power of new machines has roughly doubled every eighteen months. This trend, known as Moore's law after Gordon Moore, the co-founder of Intel who first pointed it out in 1965, is driven by miniaturization. Computer power increases as it becomes possible to pack more circuit components into a given space. But, if Moore's law is to hold fast for another twenty-five years or so, electronic devices must shrink to nanometre sizes: the scale of molecules.

No one yet knows how that will be achieved, for at such scales the silicon transistor – the workhorse of integrated circuits – becomes too leaky a switch. In order to continue making computers faster and more powerful, a growing school of thought says that their components will have to be individual molecules. This is so different a vision from conventional information technology that it would be a bold or reckless speculator who invests in it.

Yet it is not a new idea. In 1974 US chemists Mark Ratner and Ari Aviram proposed a design for a single-molecule rectifier (a device that passes current in only one direction). Just a few years later, carbon-based polymers were discovered that can conduct electricity, and researchers began to hope that individual molecules of these materials might supply the wires for a 'molecular computer'. The field of *molecular electronics* was born.

But for the next decade or so, nothing much happened. It was an idea before its time, lacking any experimental means to synthesize, arrange, or probe the kinds of molecular devices it dreamed of. In recent years, several paths have converged to create a resurgence in the field, and at last molecular electronics – and its corollary, molecular computing – are starting to gather serious attention from the people who matter: the companies who build computers.

One of the central ingredients of a molecular information-processing technology is the switch: a device to supplant the transistor. In the most rudimentary terms, a switch can exist in two different stable states – 'on' and 'off'. The transistor passes a current when 'on', and blocks it when 'off'. But a switch is useful for information processing only if it can be hooked up to others, so that the devices can talk to each other and pass information to and fro. This is hard to achieve with molecules.

Yet something of the kind was announced in 1999 by James Heath and co-workers from the University of California at Los Angeles, in collaboration with scientists from the computer giants Hewlett-Packard. They interconnected several switches based on an organic molecule to produce an electrically controlled logic gate.

In computer circuits, information is encoded in binary form – as a series of 1's and 0's. A signal of 1 corresponds to a electrical pulse at a certain voltage; a signal of 0 corresponds to zero voltage. Those are the only two signals dispatched through the circuit – there are no signals of ½ or 2. Data is encoded in the sequence of 1's and 0's just as DNA encodes information in a sequence of nucleotide bases. The binary code is simpler than the genetic code: it has only two characters. Each unit of information in the coded signal – each 1 and 0 – is called a binary digit, or bit.

Computers manipulate binary information and perform calculations using logic gates: devices or circuits that make decisions. A logic gate receives one or more input signals and releases one or more output signals. The outputs depend on what the inputs say. An AND gate, for instance, receives two input bits and produces one output. If both inputs are 1's, the output is also 1; but any other combination of inputs generates an output of 0. Simple combinations of logic gates like these can perform arithmetic: for example, reading two numbers encoded in binary form and generating an output that encodes their sum (addition) or their difference (subtraction).

Heath and colleagues constructed AND gates from a molecular switch called a rotaxane. This is an assembly of two molecules: a ring threaded on a rod. The ring is prevented from falling off by big capping units fixed to the ends of the rod. The rod is designed to attract the ring, so that the two molecules interlock spontaneously when mixed together. The end caps are added afterwards. Heath's colleague Fraser Stoddart at UCLA developed techniques for making molecular assemblies like this while at the University of Sheffield in England in the late 1980s.

The researchers arranged rotaxane molecules in a layer on a metal electrode, and deposited fine metal wires on top of them. By applying a voltage to the wires, the molecules could be switched from a low-conductivity to a high-conductivity state. Many thousands of molecules, attached to a single wire, constituted a single, switchable device. The researchers connected several of these devices together to make an AND gate.

In principle, they said, it should be possible to build each device from just a single switchable molecule. But it is difficult to make electrical connections to, and measure tiny currents through, single molecules. Yet even this is not impossible. Mark Reed, James Tour, and co-workers in the USA have measured the electrical conductivity of a single 'molecular wire' connecting two gold electrodes.

Fraser Stoddart, in collaboration with Vincenzo Balzani at Bologna and their co-workers, has demonstrated a different logic operation, called XOR, in a single-molecule gate. Like AND, an XOR gate has two inputs and one output. The output signal is 1 if the inputs are different (0,1 or 1,0) and 0 if they are the same (0,0 or 1,1). The researchers observed this behaviour in a pseudorotaxane – a threaded-ring molecular assembly with no end stoppers, so that the ring can slip off the rod.

Rather than using electrical signals, the device received chemical inputs and produced an optical output. That is to say, it altered its light-emission behaviour (fluorescence) depending on whether two 'chemical signals' were 'on' or 'off' – whether the two chemicals were present or not (Fig. 40). This is analogous to the way that cell-surface receptor proteins work (see page 119) – they send out some kind of signal depending on whether or not they have bound their target molecules.

Using similar principles, A. Prasanna de Silva and Nathan McClenaghan at the University of Belfast in Northern Ireland have combined two molecular logic gates in a way that permits them to conduct elementary arithmetic. They have, in other words, been able to use molecules to count and to perform simple sums, such as 1 + 1 = 2.

It is, of course, a long way from adding one plus one to making a computer that compares with silicon-based devices. But studies such as these demonstrate an important principle: molecules can be

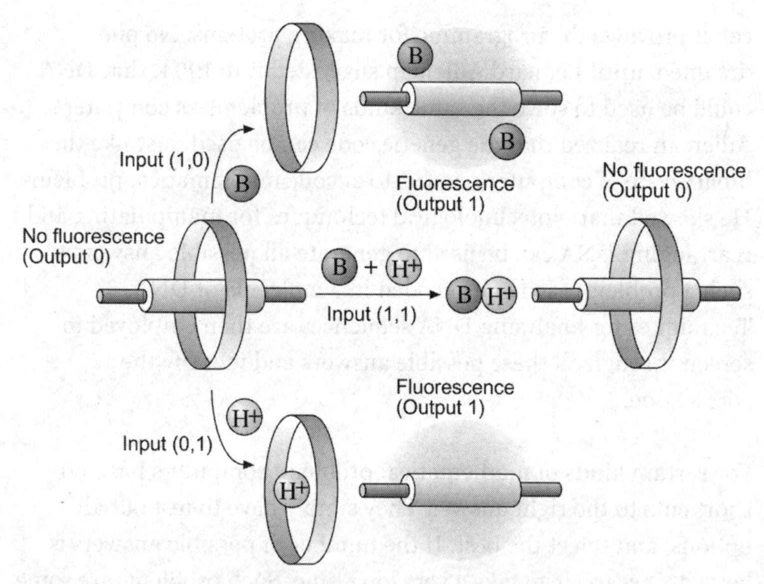

40. Molecular logic conducted by a needle-and-thread molecular assembly called a rotaxane.

used for computation, and at the level of one device per molecule. The molecular computer is at last starting to look like more than canny advertising.

The closer we look at the idea, the more we can see similarities with the challenges that the body faces: how to arrange molecules where you want them, how to transmit and amplify signals, how to grow wires between two switching devices (such as neurons), how to cope with errors, how to control the relative timing of events. The computer engineers of the future may need to know a lot of biology.

DNA computing

As if to drive that message home, in recent years some scientists have shown that computing can be conducted using DNA. This brings us full circle, for I began by suggesting that DNA provides a kind of proof-of-principle for molecular computation. But in the cell it provides the programme for making proteins. No one dreamed, until Leonard Adleman suggested it in 1994, that DNA could be used to solve the same kinds of problems as computers. Adleman realized that the genetic code can be used, just like the binary code of computer science, to encode mathematical problems. He showed that biotechnological techniques for manipulating and rearranging DNA can be used to generate all possible answers to such a problem, each one encoded in a molecule of DNA. Techniques for analysing DNA sequences are then employed to screen through all these possible answers and identify the correct one.

For certain kinds of mathematical problem, computers have no short cuts to the right answer. They simply have to test out all options, and select the best. If the number of possible answers is large, this search can take a very long time. Such problems are some of the hardest to solve using conventional computers. A classic example is the 'travelling salesman' problem, which entails working out the shortest route connecting a large number of points in space ('cities') so that each is visited just once.

Adleman showed that, by shuffling and splicing short segments of DNA at random, all the solutions to these problems may be encoded in a test tube of single-stranded DNA molecules. The number of such solutions might be huge – but the number of molecules in a test tube is greater still. And, because all the possible answers are produced and tested at once, rather than one at a time, in principle DNA computing can find the 'best' answer rapidly.

Whether or not DNA computing proves to be useful in a practical sense, it has a strong allegorical appeal. It drives home the message that the molecular basis of life is rooted in the manipulation of information. It is often said that each age tends to interpret the world through models derived from its most advanced technology, and so maybe in the Age of Information we should be wary of becoming too dogmatic about such a (partial) answer to the perennial question that haunted Haldane, Schrödinger, and countless others. It is perhaps more important that we regard this as a demonstration of the fabulously dynamic, interactive world inhabited, unseen and too often unsung, by molecules.

Notes

1. Engineers of the Invisible

1 'The sergeant beckoned the waitress'. F. O'Brien, *The Dalkey Archive* (Normal, Ill.: Dalkey Archive Press, 1993), 80.

2 'Did you ever study the Mollycule Theory'. Ibid. 81.
'Mollycules is a very intricate theorem'. Ibid. 81–2.

4 Primo Levi's famous book is *The Periodic Table* (London: Abacus, 1986).
The quotations in the box are from ibid. 3, 58, 171; the translation is copyright Shocken Books Inc. (1975).

6 John Horgan's book is *The End of Science* (London: Little, Brown & Co., 1997).
The several other recent books I refer to include the (otherwise excellent) M. Kaku, *Visions* (New York: Doubleday, 1997), and J. Maddox, *What Remains to be Discovered* (London: Macmillan, 1998).

7 Roald Hoffmann's book is *The Same and Not the Same* (New York: Columbia University Press, 1995).

8 'a "low" chemistry, almost culinary'. P. Levi and T. Regge, *Conversations* (London: I. B. Tauris & Co., 1989).
Thomas Pynchon, *Gravity's Rainbow* (London: Penguin, 1987).

9 The quotation in the box is from ibid. 249–50.

13 My book is *H_2O: A Biography of Water* (London: Weidenfeld & Nicolson, 1999).

15 'the profession I studied in school'. P. Levi, *The Monkey's Wrench* (London: Penguin, 1987), 142–3.

24 Primo Levi 'put it this way' in ibid. 144.

26 For the story of the aniline dye industry, see A. S. Travis, *The Rainbow Makers* (Bethlehem, Pa.: Lehigh University Press, 1993), and S. Garfield, *Mauve* (London: Faber & Faber, 2000).

29 'as you can imagine'. Levi, *The Monkey's Wrench*, 144.

30 The papers reporting the synthesis of taxol are K. C. Nicolaou *et al.*, 'Total Synthesis of Taxol', *Nature*, 367 (1994), 630; R. A. Holton *et al.*, "First Total Synthesis of Taxol. 1: Functionalization of the B Ring', *Journal of the American Chemical Society*, 116 (1994), 1597; R. A. Holton *et al.*, 'First Total Synthesis of Taxol. 2: Completion of the C and D Rings', *Journal of the American Chemical Society*, 116 (1994), 1599.

32 Müller's group describe their 'big wheels' in A. Müller *et al.*, 'Formation of a Ring-Shaped Reduced "Metal Oxide" with the Simple Composition $[(MoO_3)_{176}(H_2O)_{80}H_{32}]$', *Angewandte Chemie International Edition*, 37 (1988), 1220.

2. Vital Signs

33 'I am not going to answer this question'. J. B. S. Haldane, *What Is Life?* (Welwyn: Alcuin Press, 1949), 58.

'Nature is what we know'. Emily Dickinson (*c.*1863), *The Complete Poems*, ed. T. H. Johnson (London: Faber & Faber, 1975), 332.

'life is a pattern of chemical processes'. Haldane, *What Is Life?*, 61–2.

35 'the cause of most phenomena'. J. J. Berzelius (1812), *Ofversigt af djurkemiens framsteg och narvarande tillstånd: A Report to the Swedish Academy of Science, Presented on 13 August 1810*, translated into English by G. Brunnmark as *A View of the Progress and Present State of Animal Chemistry* (London, 1818), 4–8.

36 'This *power to live*'. Ibid.

45 Allan Wilson's group reported their findings in R. L. Cann, M. Stoneking, and A. C. Wilson, 'Mitochondrial DNA and Human Evolution', *Nature*, 325 (1987)/31. This article was much criticized on methodological grounds: see H. Gee, *Nature*, 355 (1992), 583.

48 'Advances in directed evolution'. J. W. Szostak, D. P. Bartel, and P. L. Luisi, 'Synthesizing Life', *Nature*, 409 (2001), 387.

3. Take the Strain

51 Space tethers are described in R. L. Forward and R. P. Hoyt, 'Space Tethers', *Scientific American*, 280/2 (February 1999), 86. For the broader picture on molecular materials, see P. Ball, *Made to Measure: New Materials for the 21st Century* (Princeton: Princeton University Press, 1997).

56 The many varieties of spider silk are discussed in F. Vollrath, 'Spider Webs and Silks', *Scientific American* (March 1992), 70.

63 Making materials from artificial genes is described in D. A. Tirrell, M. J. Fournier, and T. L. Mason, 'Genetic Engineering of Polymeric Materials', *MRS Bulletin* (July 1991), 23.

66 The discovery of carbon nanotubes was reported in S. Iijima, 'Helical Microtubules of Graphitic Carbon', *Nature*, 354 (1991), 56.
For the story of the discovery of the fullerenes, see J. Baggott, *Perfect Symmetry* (Oxford: Oxford University Press, 1994).

69 The tentative step towards the production of nanotube fibres is described in B. Vigolo *et al.*, 'Macroscopic Fibers and Ribbons of Oriented Carbon Nanotubes', *Science*, 290 (2000), 1331.

4. The Burning Issue

73 The Second Law of Thermodynamics is discussed in depth in P. W. Atkins, *The Second Law* (New York: W. H. Freeman & Co., 1984).

74 Erwin Schrödinger, *What Is Life?* (Cambridge: Cambridge University Press, 1944).

90 The work on artificial mimics of photosynthesis is described in G. Steinberg-Yfrach *et al.*, 'Conversion of Light Energy to Proton Potential in Liposomes by Artificial Photosynthetic Reaction Centres', *Nature*, 385 (1997), 239; G. Steinberg-Yfrach *et al.*, 'Light-Driven Production of ATP Catalysed by F_0F_1–ATP Synthase in an Artificial Photosynthetic Membrane', *Nature*, 392 (1998), 479.

91 '$CH_3C_6H_2(NO_2)_3 + Hg(CNO)_2$'. A. Huxley, *Brave New World* (London: Granada, 1977), 48.

92 The synthesis of explosive octanitrocubane is described in M.-X. Zhang, P. E. Eaton, and R. Gilardi, 'Hepta- and Octanitrocubanes', *Angewandte Chemie International Edition*, 39 (2000), 401.

5. Good Little Movers

94 Richard Feynman's article 'There's Plenty of Room at the Bottom' appeared in *Engineering and Science* magazine (February 1960). More accessible versions are available in B. C. Crandall and J. Lewis (eds.), *Nanotechnology: Research and Perspectives* (Cambridge, Mass.: MIT Press, 1992), or at *http://www.zyvex.com/nanotech/feynman.html.*

96 'we don't have those tweezers'. P. Levi, *The Monkey's Wrench* (London: Penguin, 1987), 144.

98 I wish I could point to a good general reference source for nanotechnology, but that is surprisingly difficult. K. Eric Drexler has done most to popularize the subject, albeit in a manner that owes as much to science fiction as to current science – see his *Engines of Creation* (London: Fourth Estate, 1990), and also K. E. Drexler, C. Peterson, and G. Pergamit (1994), *Unbounding the Future* (Reading, Mass.: Addison-Wesley, 1994). And, for critical comment on Drexler's vision of nanotechnology, see G. Stix, 'Trends in Nanotechnology: Waiting for Breakthroughs', *Scientific American* (April 1996), 94; and D. E. H. Jones, 'Technical Boundless Optimism', *Nature*, 374 (1995), 835. Ed Regis's *Nano!* (New York: Bantam Books, 1995), is a readable account, but centres around Drexler's vision. The volume by Crandall and Lewis (above) has some nice contributions, if slightly technical and now outdated. Michael Gross's *Travels to the Nanoworld* (New York: Plenum, 1999), is well worth reading but sometimes idiosyncratic in its choices. The web site of Drexler's Foresight Institute (*http://www.foresight.org*) does a good job of collating up-to-date popular and technical articles.

106 The work of Kinosita's group is described in Y. Arai *et al.*, 'Tying a Molecular Knot with Optical Tweezers', *Nature*, 399 (1999), 446.

107 The artificial molecular motor is described in N. Koumura, R. W. J. Zijlstra, R. A. van Delden, N. Harada, and B. L. Feringa, 'Light-Driven Monodirectional Molecular Motor', *Nature*, 401 (1999), 152.

108 Ross Kelly's rotary motor is described in T. R. Kelly, H. De Silva, and R. A. Silva, 'Unidirectional Rotary Motion in a Molecular System', *Nature*, 401 (1999), 150.

110 The field of synthetic molecular motors and machines is reviewed in V. Balzani, A. Credi, F. M. Raymo, and J. F. Stoddart, 'Artificial Molecular Machines', *Angewandte Chemie International Edition*, 39 (2000), 3348–91.

110 The work of Stan Leibler's group is described in F. J. Nédélec, T. Surrey, A. C. Maggs, and S. Leibler, 'Self-Organization of Microtubules and Motors', *Nature*, 389 (1997), 305.

111 Viola Vogel and colleagues reported their work in J. Dennis, J. Howard, and V. Vogel, 'Molecular Shuttles: Directed Motion of Microtubules along Nanoscale Kinesin Tracks', *Nanotechnology*, 10 (1999), 232.
The ATP synthase rotors made by Carlo Montemagno and colleagues are described in R. K. Soong *et al.*, 'Powering an Inorganic Nanodevice with a Biomolecular Motor', *Science*, 290 (2000), 1555–8.

6. Delivering the Message

115 The endocrine system and other systems of biomolecular communication are discussed in S. Aldridge, *Magic Molecules: How Drugs Work* (Cambridge: Cambridge University Press, 1998).

126 For an excellent discussion of neurotoxins and mind-altering drugs, see J. Mann, *Murder, Magic, and Medicine* (Oxford: Oxford University Press, 1992).

129 For an overview of supramolecular chemistry, see J.-M. Lehn, *Supramolecular Chemistry* (Weinheim: VCH, 1995). A less technical introduction is given in J.-M. Lehn and P. Ball, 'Supramolecular Chemistry', in N. Hall (ed.), *The New Chemistry* (Cambridge: Cambridge University Press, 2000).

130 The 'signal-transducing' synthetic molecular receptor is described in R. Krauss, H.-G. Weinig, M. Seydack, J. Bendig, and U. Koert, 'Molecular Signal Transduction through Conformational Transmission of a Perhydroanthracene Transducer', *Angewandte Chemie International Edition*, 39 (2000), 1835.

7. The Chemical Computer

133 John Hopfield speaks about biological existence theorems in the text of a talk entitled 'Biological information processing and molecular nanocomputing', given at the *Nature* conference on nanotechnology, Tokyo, January 1992.

137 The work of Leslie Orgel and others on replicating molecules is discussed in L. Orgel, 'Molecular Replication', *Nature*, 358 (1992), 203.

138 Günter von Kiedrowski reported his molecular replicators in G. von Kiedrowski, 'A Self-Replicating Hexadeoxynucleotide', *Angewandte Chemie International Edition*, 25 (1986), 932.

143 Nadrian Seeman's DNA cages are described in J. Chen and N. C. Seeman, 'Synthesis from DNA of a Molecule with the Connectivity of a Cube', *Nature*, 350 (1991), 631; Y. Zhang and N. C. Seeman, 'The Construction of a DNA Truncated Octahedron', *Journal of the American Chemical Society*, 116 (1994), 1661.

144 Chad Mirkin and colleagues reported their DNA-based method of nanoparticle assembly in C. A. Mirkin, R. L. Letsinger, R. C. Mucic, and J. J. Storhoff, 'A DNA-Based Method for Rationally Organizing Nanoparticles into Macroscopic Materials', *Nature*, 382 (1996), 607.

145 The peptides that recognize different semiconductor surfaces are described in S. R. Whaley, D. S. English, E. L. Hu, P. F. Barbara, and A. M. Belcher, 'Selection of Peptides with Semiconductor Binding Specificity for Directed Nanocrystal Assembly', *Nature*, 405 (2000), 665.

146 The molecule-based logic devices built by Jim Heath and co-workers are described in C. P. Collier *et al.*, 'Electronically

Configurable Molecular-Based Logic Gates', *Science*, 285 (1999), 391.

148 The rotaxane-based logic device is reported in A. Credi, V. Balzani, S. J. Langford, and J. F. Stoddart, 'Logic Operations at the Molecular Level: An XOR Gate Based on a Molecular Machine', *Journal of the American Chemical Society*, 119 (1997), 2679. The measurement of the electrical conductivities of single molecules is described in M. A. Reed, C. Zhou, C. J. Muller, T. P. Burgin, and J. M. Tour, 'Conductance of a Molecular Junction', *Science*, 278 (1997), 252; M. A. Reed and J. M. Tour, 'Computing with Molecules', *Scientific American* (June 2000), 86. Molecules that perform arithmetic are described in A. P. de Silva and N. D. McClenaghan, 'Proof-of-Principle of Molecular-Scale Arithmetic', *Journal of the American Chemical Society*, 122 (2000), 3965.

150 Leonard Adleman's proposal for DNA computing is described in L. M. Adleman, 'Molecular Computation of Solutions to Combinatorial Problems', *Science*, 266 (1994), 1021; L. M. Adleman, 'Computing with DNA', *Scientific American* (August 1994), 54.

Further reading

Atkins, P. W. *Atoms, Electrons, and Change* (New York: W. H. Freeman & Co., 1991).

— *Molecules* (New York: W. H. Freeman & Co., 1991).

Ball, P., *Designing the Molecular World: Chemistry at the Frontier* (Princeton: Princeton University Press).

Emsley, J. *Molecules at an Exhibition* (Oxford: Oxford University Press, 1998).

Goodsell, D., *The Machinery of Life* (New York: Springer-Verlag, 1993).

— *Our Molecular Nature: The Body's Motors, Machines and Messages* (New York: Springer-Verlag, 1996).

Hall, N. (ed.), *The Age of the Molecule* (London: Royal Society of Chemistry, 1999).

— (ed.), *The New Chemistry* (Cambridge: Cambridge University Press, 2000).

Hoffmann, R., *The Same and Not the Same* (New York: Columbia University Press, 1995).

“牛津通识读本”已出书目

古典哲学的趣味
人生的意义
文学理论入门
大众经济学
历史之源
设计，无处不在
生活中的心理学
政治的历史与边界
哲学的思与惑
资本主义
美国总统制
海德格尔
我们时代的伦理学
卡夫卡是谁
考古学的过去与未来
天文学简史
社会学的意识
康德
尼采
亚里士多德的世界
西方艺术新论
全球化面面观
简明逻辑学
法哲学：价值与事实
政治哲学与幸福根基
选择理论
后殖民主义与世界格局

福柯
缤纷的语言学
达达和超现实主义
佛学概论
维特根斯坦与哲学
科学哲学
印度哲学祛魅
克尔凯郭尔
科学革命
广告
数学
叔本华
笛卡尔
基督教神学
犹太人与犹太教
现代日本
罗兰·巴特
马基雅维里
全球经济史
进化
性存在
量子理论
牛顿新传
国际移民
哈贝马斯
医学伦理
黑格尔

地球
记忆
法律
中国文学
托克维尔
休谟
分子
法国大革命
民族主义
科幻作品
罗素
美国政党与选举
美国最高法院
纪录片
大萧条与罗斯福新政
领导力
无神论
罗马共和国
美国国会
民主
英格兰文学
现代主义
网络
自闭症
德里达
浪漫主义
批判理论

德国文学
戏剧
腐败
医事法
癌症
植物
法语文学
微观经济学
湖泊
拜占庭
司法心理学
发展
农业
特洛伊战争
巴比伦尼亚
河流

儿童心理学
时装
现代拉丁美洲文学
卢梭
隐私
电影音乐
抑郁症
传染病
希腊化时代
知识
环境伦理学
美国革命
元素周期表
人口学
社会心理学
动物

电影
俄罗斯文学
古典文学
大数据
洛克
幸福
免疫系统
银行学
景观设计学
神圣罗马帝国
大流行病
亚历山大大帝
气候
第二次世界大战
中世纪